FIREFLY

STARGAZING WITH BINOCULARS

ROBIN SCAGELL · DAVID FRYDMAN

A FIREFLY BOOK

Published by Firefly Books Ltd. 2008

First printing

Publisher Cataloging-in-Publication Data (U.S.)

Scagell, Robin.
Stargazing with binoculars / Robin Scagell and David Frydman.
[208] p. : col. photos., ill., maps ; cm.
Includes index.
Summary: Guide to stargazing with binoculars in both northern and southern hemisphere, with comprehensive listings of types of binoculars, mounts and other accessories, and advice on what objects can be seen, and when, and how to find them.
ISBN-13: 978-1-55407-368-9 (pbk.)
ISBN-10: 1-55407-368-5 (pbk.)
1. Astronomy -- Observers' manuals. 2. Stars -- Observers' manuals. 3. Binoculars. I. Frydman, David. II. Title.
523.80223 dc22 QB63.S334 2008

Library and Archives Canada Cataloguing in Publication

Scagell, Robin
Stargazing with binoculars / Robin Scagell and David Frydman.
Includes index.
ISBN-13: 978-1-55407-368-9 (pbk.)
ISBN-10: 1-55407-368-5 (pbk.)
1. Astronomy--Observers' manuals. 2. Binoculars. 3. Stars-- Observers' manuals. 4. Constellations--Observers' manuals.
I. Frydman, David II. Title.
QB63.S366 2008 522 C2007-906741-7

Published in the United States by
Firefly Books (U.S.) Inc.
P.O. Box 1338, Ellicott Station
Buffalo, New York 14205

Published in Canada by
Firefly Books Ltd.
66 Leek Crescent
Richmond Hill, Ontario L4B 1H1

Published in Great Britain by Philip's
a division of Octopus Publishing Group Ltd,
2-4 Heron Quays, London E14 4JP
An Hachette Livre UK Company

Printed in China

Title page:
Robin Scagell/Galaxy

CONTENTS

1 · INTRODUCTION

Every stargazer has binoculars. Far from being the poor person's telescope, binoculars have their own special advantages. Even the most advanced amateur astronomers, whose telescopes cost more than their automobiles, own binoculars, and regularly put them to good use.

So what can binoculars do for you that a telescope can't? Their big advantage is that they give you a much wider view of the sky than most telescopes. Telescopes are essential for giving close-up views of objects, such as planets. But for general stargazing, binoculars show you the big picture.

Imagine yourself out under the stars, on a beautiful dark clear night somewhere deep in the country. The immensity of the heavens stretches above you. There are stars from horizon to horizon. As you gaze upward, you see little knots of stars and vague misty patches. What are they in reality? You raise your binoculars and suddenly those little stars are spread out before your eyes as a mass of glittering points of light.

The little misty patches are transformed, too. Though they remain as gray misty blobs, they now take on some real shape and in some cases we can see intricate detail in stellar birthplaces.

Binoculars can see much farther. With the unaided eye you can see our nearest large galaxy, in Andromeda. But with binoculars in a good dark sky you can start to pick out galaxies in the Virgo cluster, over 20 times farther away.

Though you can really only see planetary detail with a telescope, there are Solar System bodies that look much better in binoculars—comets being a good example. When a good comet comes along—though this is a fairly rare event—binoculars are perfect for picking out its full extent, which may cover several degrees of sky. Even run-of-the-mill comets, of which there are usually one or two a year, are good targets for binoculars.

The wide field of binoculars and their portability make them the instrument of choice on numerous occasions. Say you want to pick up some elusive cluster or nebula, a bright asteroid such as Ceres, or a recently discovered comet. Yes, you can set up a computer-controlled telescope which could guide you straight to the spot—if you have aligned it correctly, and if there is an adequate power supply, both of which can be quite big Ifs at times. But with binoculars and a half-decent star chart you can go to the object straight away, with a little practice. Even owners of good telescopes may take a look first with binoculars, just to check that the object isn't behind a tree, for example.

Suburbanites and citydwellers will find binoculars even more useful than their country cousins. In skies where you can see hardly any stars with the naked eye, binoculars will help you to pick out stars and asterisms that would otherwise remain invisible. A small constellation such as Cancer or Delphinus may be revealed to you for the first time.

And the best thing is that these wonderful instruments need not cost you very much. Though it is usually best to avoid the very cheapest, with a bit of care you can get perfectly serviceable binoculars for the price of a modest cellphone, and from time to time there are real bargains available. In fact, as we shall see, it is not always a good idea to pay top money for your binoculars. You don't necessarily need the ruggedness of binoculars designed for marine use, or for birdwatching, if the worst that they are likely to suffer is a bit of dew on the lenses.

Why binoculars?

You can find out more about the choice available in *Chapter Five*, but to start with you need only know a little about binoculars. Basically,

▲ *The Sun has set and the stars are beginning to appear—giving the promise of a good night's observing ahead using nothing more than binoculars.*

binoculars are two low-power, wide-field telescopes side by side, with their eyepieces at the same distance apart as your eyes. Whereas with a similar specification of telescope you might have to squint a bit to exclude the direct view, with binoculars you are able to look through both eyes simultaneously. This has several advantages. We are accustomed to using both eyes, so the view feels more natural to us. Nearby objects can be seen in three dimensions, as the natural stereo vision is enhanced especially if you are using the classic design of binoculars in which the two main objective lenses are spaced more widely than your own eyes.

Although astronomical objects are all much too far away for there to be any stereo effect, using both eyes has a great advantage when it comes to seeing detail. When you are trying to see a faint object, or pick out a faint star, you are usually at the limits of your eyesight. Using both eyes instead of one can be a great help. Some people even go to the extent of building giant binoculars—basically, two powerful telescopes—in order to get better views of astronomical objects, while others prefer to use binocular eyepieces on their telescopes, which split the light into two eyepieces. Even though the brightness of the view is considerably reduced, the ability to use both eyes gives them an advantage.

Binoculars are specified by their magnification and by the size of the main lenses in millimeters—such as 8 × 40, pronounced "8 by 40." In this case, the magnification is 8, while the main lenses (the objective lenses) are 40 mm (1.6 in) in diameter. Put simply, the larger the objective lenses, the brighter the image, but the heavier the binoculars. You might imagine that higher-magnification binoculars would be much bigger than those of low magnification, but the magnification has nothing to do with their bulk. Look at 7 × 50 and 10 × 50 binoculars of the same design, and you will see that they look pretty well identical. However, increasing the magnification means that the light from the objective lenses is spread over a greater area, giving a dimmer view than at lower magnifications. So in practice the higher-magnification binoculars are usually made in larger sizes to avoid very dim images.

Incidentally, it is not uncommon to hear people talking about "a pair of binoculars." Strictly speaking this is wrong. The "bi"- prefix in binoculars means "two," so really we should speak in the singular about "a binocular" just as we refer to "a bicycle." Here we will refer to binoculars in the plural only, which is the accepted usage. There are devices called monoculars, which are half binoculars, for one eye only. They have the advantage of portability, but of course lack the benefits of using both eyes. Because they are less popular, there is no great price advantage in using a monocular.

Using binoculars for astronomy

With few exceptions, pretty well all binoculars have some use in astronomy. This means that you can use virtually any general purpose binoculars for stargazing, whether you bought them for birdwatching, racing, yachting, or just plain gazing at the landscape when you stop at a viewpoint, which is probably the most common use of all. Even cheap binoculars will be of use, as long as they work as they should, though a better instrument may well give you an improved view.

If possible, your first look should be at the Moon, which, since it is so easy to find and focus on, is one of the most rewarding objects to study, and is unmistakable. This might seem odd, but many people do have problems to start with when viewing stars and constellations. It can be hard to work out exactly which bit of the sky you are looking at, and stars are not the easiest thing to focus on if they are not fairly sharp to start with. With the Moon there is no doubt, and you can immediately appreciate the difference in scale of view that the binoculars make.

A new world appears before your eyes. While you can probably see the most obvious dark markings on the Moon with the naked eye, through binoculars it becomes a more obviously rounded body. You start to see mountains and a few craters, particularly toward the shadow line (the terminator). Perhaps the best time to look is around first or last quarter, when the terminator is more or less central and bisects the disc. The craters are particularly obvious along its line. And the worst time to look is near full Moon, both because there are fewer shadows to give some sense of the relief of the landscape, and also because it is often so bright that you can be momentarily dazzled. Although you won't do your eyes any harm at all by looking at the full Moon, it can be decidedly uncomfortable if your eyes have become adapted to darkness, and it is likely that you will have a spot in front of your eyes for several minutes.

▲ *Even the simplest binoculars will reveal a wealth of detail on the Moon, and you can spend a long time studying our neighbor in space.*

▲ *Visual versus camera. The Whirlpool Galaxy, M51, as it appears in the sky. The bright double star at the top is Mizar, and that at center left is Alkaid. These are the two stars at the end of Ursa Major's tail. This view covers 13½° of sky from top to bottom, so binoculars will show only the central half or third of the area, and will not reveal so many stars.*

▲ *M51 as photographed using a long exposure with a 200 mm (7.8 in) telescope by Nick King from Harrow, just 19 km (12 mi) from Central London. Light pollution from this site makes it hard to see this galaxy at all using binoculars, but computer processing of the image has brought out fine detail in the spiral arms of the galaxy.*

While looking at the Moon, practice raising the binoculars to your eyes and finding the Moon, and putting them down and taking a look directly a few times. This might seem a little pointless, but it gives you an idea of the limitations of the field of view, and also of the scale of the binocular field of view. This will help a bit when you are searching for a particular object in a sky where there are few bright stars to act as landmarks.

Also while looking at the Moon you can begin to assess the quality of the image. Astronomical objects can often be a severe test of the optical quality of an instrument. The high contrast of the bright Moon will show up any stray reflections in the image, caused by light being reflected off its internal surfaces. By day, every part of the image is usually bright, so particular reflections become merged with the rest of the view. But if the dark sky around the Moon has a flare at some point which changes position relative to the Moon as you move it around the field of view, you can be sure that it is a defect of the instrument.

Having looked at the Moon, take a look at any bright stars that are around. You may be disappointed if you are expecting them to show any sort of detail, but in fact all stars are so far away that virtually no telescope will show them as anything other than points of light. While the image should be genuinely pointlike, a bright star can be so dazzling that it can appear to have rays coming off it, making it sparkle. This is caused by defects of eyesight, which you would get when looking at bright objects directly in any case. When we think of the appearance of a star, we usually give it points—just as in the classic Christmas card appearance, or indeed the stylistic five-pointed star. This is entirely due to the defects of eyesight, from which we all suffer.

But as well as the bright star, you will probably see other, fainter, stars nearby, which are invisible when you look directly. These are fainter stars that are being revealed by the binoculars. They are usually not so bright that they dazzle the eye, and they should appear much more as points of light. So your disappointment at not seeing a great magnified globe of gas is tempered by the unexpected sight of it being surrounded by a host of smaller jewels. Just stargazing with binoculars on a good night can be tremendously rewarding, even if you don't know the objects you are looking at. Photographs in books do not do justice to the delicate appearance of a starry field of view in binoculars.

With your first views of the heavens comes a realization that the binoculars which seemed so light and handy when taking a quick look at a distant boat on the sea or some feature of the landscape now become considerably heavier. Your arms soon start to ache, which reduces your viewing comfort. Looking upward at an angle places a greater strain on you than looking horizontally, so always try to find some way of supporting your arms when viewing the sky. However this is not at all easy when trying to view objects that are more or less overhead. In *Chapter Six* you will find information about devices that can improve your viewing experience, including binocular stands, tripods, and mirror viewers.

You will also find that it becomes hard to hold the binoculars steady. When you are trying to see fine detail, the slightest movement of your arms becomes magnified several times, making the job much harder. Again, support is very helpful. But these days, technology has come to the rescue with image-stabilizing binoculars. These usually contain sensors that detect unsteady movement and then move one of the optical components to compensate. The difference can be striking in use.

Gaze at your chosen star and it is all over the place as your arms move. Press a button and the movement all but stops. The system does allow you to move the binoculars from one spot to another without trying to lock on to objects, and it is most sensitive to the sort of natural movements that you want to counteract.

Such binoculars are costly, though in their favor they often use good-quality optical systems as well, and so are particularly well suited to astronomy. A cheaper alternative is to fix your binoculars to a tripod, and these days many instruments have a bush that will allow you to attach them to a suitable photographic tripod, usually via an adaptor which may have to be bought separately. This is discussed further on page 181.

Expand your stargazing by picking interesting-looking groups of close stars and noticing how their appearance changes when you bring binoculars to bear on them. Both their brightness and the scale of view change, and it is therefore a good idea to get some practice in, particu-

▲ The Milky Way—here, in Perseus and Cassiopeia—is a great area for just plain stargazing. This wide-field view shows stars down to magnitude 9.

larly if you are using higher-power binoculars such as 12 × 50s. Even experienced binocular users occasionally find that the object they want to look at can be stubbornly in the wrong place when they try to observe it through binoculars, particularly when it is high up. What looks faint to the naked eye can appear bright through binoculars—so have you actually found the right star? Only practice will tell you.

Finally, a word of warning. If you are expecting to see the same sights as those that appear in photographs you will probably be disappointed. Most deep sky objects (that is, nebulae, galaxies, and so on) are fairly dim, even in good binoculars or a telescope. The human eye is not very sensitive to color at these low light levels, so the vivid colors of nebulae are just not visible. It is also necessary to emphasize the brightness of objects so that they will show up well in an illustration. Typically, an illustration will make the brightest part of a nebula appear quite light on the page, whereas at the eyepiece it might be just a pale glow. If a book were to be illustrated with total accuracy, you would see nothing but mostly black rectangles.

However, if the drama and color are lacking in the binocular sky, other aspects more than make up for it. For one thing, you are actually seeing these objects with your own eyes: the light which you are seeing is the same light that left the object maybe thousands or millions of years ago. Though the tiny soft glow of a galaxy in binoculars may not have the same visual impact as a stunning photograph, it is something that you have hunted down for yourself, armed with nothing more than the binoculars and your own knowledge of the sky. Most observers will return to the same object time and time again, and appreciate it for what it is.

And the other side of the coin is that some objects look far more spectacular through binoculars than they do on the printed page. The Moon is one, as the sheer brilliance and contrast of the real thing just can't be matched on paper. Star clusters can also be breathtaking. Photographs can only represent stars as blobs of varying sizes. In reality, the brightest stars appear as dazzling points of light, which are much more appealing than large blobs.

In this book we have used a combination of photographs that approximate to the binocular view, long-exposure photographs taken through larger instruments that reveal more than you can possibly see, and drawings made at the eyepiece. Even the photographs that try to show the same view as binoculars do not always give the right impression—they generally show many more stars than you can actually see, in order to make the fainter objects easily visible. But the aim is to help you to locate the objects, and to know in advance whether it is large or small, bright, or faint. There is no substitute for looking for yourself.

2 · LEARNING THE SKY

If you want to locate the most interesting objects in the sky, you first have to learn at least some of the constellations (star patterns). Fortunately, this is not as hard as you might think. The key is to pick out a few constellations each month, and then using those as landmarks you will be able to find your way to the fainter features.

The main obstacle to learning the sky is that it changes from hour to hour and from month to month, also changing with your location on the Earth's surface. But the changes are slow and progressive, so the stars from one month are still around next month, but shifted across the sky.

To see how the sky moves, look at the miniature sky maps opposite and overleaf, which show what happens to just a few key star groups as time goes by and as you change location. You can find more detailed views on subsequent pages. To make things simpler, we have chosen two locations that will work for the majority of readers. The northern hemisphere maps are for latitude 45°N, and are suitable for most of the UK, Europe, and much of North America. Those for the southern hemisphere are for latitude 35°S, which accommodates the key cities of Australia and South Africa. Between them they represent the whole sky, so from virtually all other latitudes you should find points of recognition.

The sky's movements are easiest to understand if you look out at the same time each night—say 10.30 pm—and if in each case, you look due south from the northern hemisphere, or due north in the southern hemisphere. Astronomers refer to the north–south line from any location as the meridian, and it can help to imagine the star groups crossing this line, like horses past a winning post.

One of the easiest constellations to recognize is Orion, so we have used the sky from the northern hemisphere in mid January at 10.30 pm as an example. The distinctive shape of Orion, with three stars in a line and four others surrounding them, is picked out with yellow lines. Looking south from latitude 45°N, you see Orion roughly halfway up the sky. If you watch for long enough, you will see that as the night draws on it moves to the right (the west). Six hours later, at 4.30 am, it is setting in the west, as seen in the next view, and other stars have moved to take its place, notably Leo, the Lion, also shown in yellow. In general stars always rise in the eastern sky and set in the western sky.

The whole sky appears to rotate once every 24 hours, as a result of the turning Earth, but because the Earth is also going round the Sun, over the course of a year we also get to see the whole sky visible from our location. Looking at 10.30 pm the next night, you will see virtually no difference, but in fact the stars arrive at the same position in the sky four minutes earlier each night. Over a couple of weeks the difference is noticeable. After three months of observing at 10.30 pm each night, Orion has moved well to the west, and is setting—the same view as for 4.30 am in January.

▲ *The sky looking south from latitude 45°N in January at 10.30 pm, with Orion high in the sky and due south.*

Over the course of a year of observing at 10.30 pm each night you will see the whole sky visible from your location. In April, Leo is due south, but by July it is setting and in its place, but lower in the sky, is another distinctive group, Scorpius, the Scorpion. In October the situation has changed again and the Square of Pegasus is in mid sky. Finally, next January we are back with Orion.

That is the sky looking south, but what if you turn round and look north? In this direction, rather than changing completely from hour to hour or from month to month, the sky simply rotates counterclockwise about its axis, which happens to point in the northern hemisphere to a moderately bright star, Polaris. So the same stars remain visible, but their orientation changes. You should always be able to spot the two key constellations, Ursa Major and Cassiopeia. The seven bright stars of Ursa Major are better known in North America as the Big Dipper.

Returning to the sky at 10.30 pm from 45°N in January, the next thing is to see what happens if we move north or south.

▲ *The sky looking south from latitude 45°N in January at 4.30 am. Orion is setting, and Leo is now in mid sky.*

▲ *By July, Leo is setting in the west and Scorpius is due south at 10.30 pm as seen from latitude 45°N.*

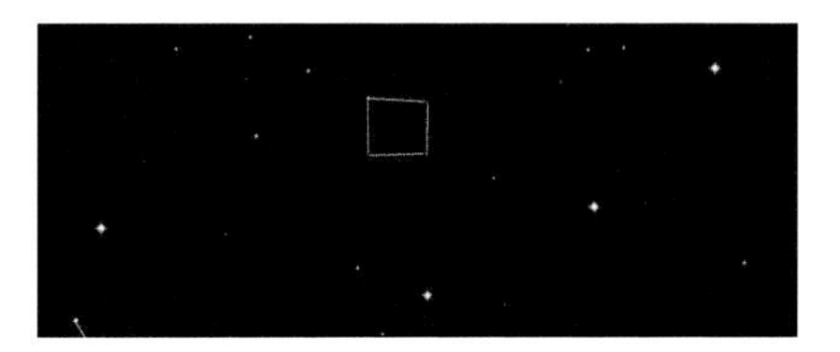

▶ *Three months later, in October, the Square of Pegasus is in mid sky. Scorpius has set, but Orion is just about to rise in the east.*

▲ *Looking south in January at 45°N. Orion is in mid sky with the stars of Canis Major to its lower left.*

▲ *From farther north, at 55°N, Orion is lower in the sky. Some stars in Canis Major are below the horizon.*

▲ *Farther south, at 30°N, Orion is higher up and more stars are visible to its south.*

▲ *From the southern hemisphere, at 35°S, Orion appears upside down compared with the northern view.*

Moving north to latitude 55°N, central Canada or northern Europe, say, we see that Orion has slipped 10° down in the sky. And from latitude 30°N, the Canary Islands or California, Orion is higher in the sky and other stars are visible below it, notably brilliant Canopus.

Moving farther south still, down to the southern hemisphere, at latitude 35°S Orion is still prominent in the mid sky, but you need to look north to see it and it is now upside down compared with the view from the northern hemisphere. Stars that were above it as seen from the northern hemisphere are now skimming the northern horizon, while some stars that were forever below the horizon from Europe or North America are now high in the sky.

If we now look to the southern sky, completely different constellations wheel around a point in the mid sky. The Southern Cross (Crux) is the landmark here, and the two hazy Magellanic Clouds are also visible in reasonably dark skies. While stars still rise in the eastern sky and set in the west, they now turn clockwise when looking south, and reach their highest looking north.

A sense of scale

One major problem with looking at star maps is that you have to convert what you see on the page into the appearance of the whole sky. The maps on the following pages, like those shown here, actually cover a large amount of sky. The distance from the bottom to the top of each map is equivalent to the horizon to a point directly overhead (known as the zenith). We are accustomed to seeing photographs that cover only a small part of the view in front of us, and when we look at a scene we really only take in the central part of it. If you photograph the sky with a camera, you often find that quite a wide-angle lens is needed to get even a single constellation in one shot. But when you are learning the sky, you really need to see how one constellation fits into the big picture. So appreciating the scale of the maps is all-important.

Returning to Orion, compare the photograph below with the maps opposite upper left and on page 20. It was taken in North Carolina at the end of December at about 10 pm. From top to bottom it covers 72° of sky, which is 80% of the distance between horizon and zenith. It gives a good idea of how much sky you actually take in at any one time.

► A wide-angle view of the December sky from North Carolina. Orion is just above center, with Sirius and Canis Major below and to the left.

▲ *When near the horizon, the Moon appears much larger than when it is high in the sky.*

But when you actually see Orion in the sky, it will probably appear larger than the photo or maps imply, particularly if it happens to be fairly low down at the time.

Our perception of size is rarely a matter of direct measurement. If you see a large building very far away, you know that it is huge even though it appears small in your field of view, and even if you can't see any other buildings at the same distance. In the case of the Sun and Moon, and star groups, when they are close to the horizon we perceive them as much larger than when they are high up. A wide-bodied airplane in the sky appears small because we have nothing to compare it with, and so do the heavenly bodies. This is why it is hard to gain an idea of the scale of star maps; this is something that you have to work out for yourself.

Because of the lack of reference points, astronomers often refer to sky measurements in degrees, with the distance from horizon to zenith being 90°. The separation between the fingers of your outstretched hand at arm's length is about 20°, and a finger at arm's length is usually more than 1°. The Sun and Moon are both ½° across. At the end of the book are some useful diagrams that give the separations in degrees of some well known stars.

You sometimes hear non-astronomers trying to describe sky distances in everyday measurements—"It's about six inches to the left of that bright star," is fairly typical. They know what they mean, but there is no real rhyme or reason to this. They are probably thinking of a ruler held up against the sky, but it is better to get some idea of degrees under your belt. In particular it is useful to know the field of view of your binoculars in degrees. Typical values range between about 4° for inexpensive 10 × 50s, and over 6° for well-made or lower-power binoculars. There is more about this, and how to work out the field of view of your own binoculars, in *Chapter Six*.

Solar System objects

What about the Sun, Moon, and planets? These are carried round by the daily and annual rotation, and can be seen equally from both hemispheres in the same part of the sky on any particular date. The difference is that while for the people of the northern hemisphere the Sun and Moon rise in the eastern sky and move upward and to the right, as seen from the southern hemisphere they move upward and to the left. So the angle at which the Sun and Moon rise or set is reversed.

The phase of the Moon is the same the world over on any particular day, though it will be slightly different as seen from Australia compared with that seen from America or Britain because of the time difference. But taking this into account, someone looking at the full Moon rising from the UK could well be seeing it at precisely the same moment as a friend in Australia is seeing it set, 10 hours later their time.

The positions of Solar System objects can't be shown on star maps because they move all the time. Tables of their positions are given in *Chapter Four.*

Using the star maps

Monthly star maps for the northern and southern hemispheres are shown from page 20 on. They are intended to get you started in finding your way around the sky, and you will need them to find the constellations that contain the interesting objects mentioned in *Chapter Three.*

Simply choose the month, then the maps for your hemisphere. The left-hand page shows the view looking away from the pole, and the right-hand page shows the view looking toward the pole in each case. Imagine each map covering half the sky and more or less surrounding you, as shown in the diagram. There is some overlap between the extreme edges of the northern and southern views, but because we are

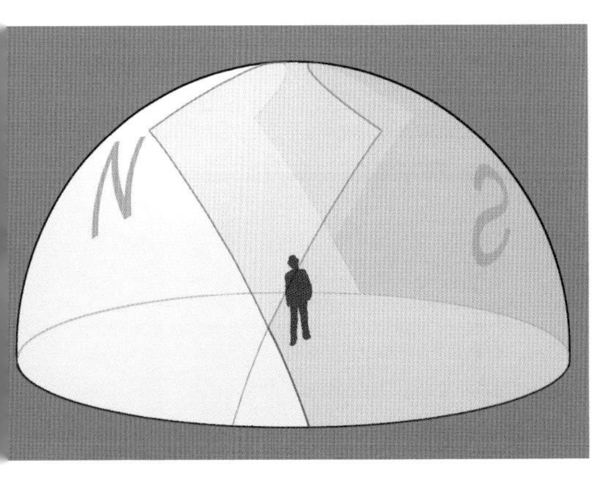

◀ *The northern and southern view maps cover almost the whole sky at the month and time stated. Each one covers a wide area of sky.*

showing the curved sky on a flat page, there is inevitable distortion in these directions.

The aim of the maps is to get you started. As you view month by month, you will see the constellations wheeling majestically before your eyes, and will start to recognize them from night to night. After less than a year, old friends will start to reappear.

Just a few constellations are picked out with lines on each map. These are intended to help you spot the patterns. On small-scale maps it can be confusing to see too many lines, which can get in the way of the stars, so we have kept them to a minimum. There is nothing official about the lines, which vary from atlas to atlas, and indeed most advanced star maps have no constellation lines at all.

About constellations

Some of the constellations or star patterns that we use today date back so many thousands of years that their origin is lost. An ancient Babylonian astronomer would have pointed out Leo or Scorpius, and would have used essentially the same names for them, though other civilizations saw different shapes in the sky. The patterns themselves have hardly changed over the intervening millennia: though the stars do move through space at a large velocity, they are so distant that their patterns have remained virtually the same throughout mankind's development.

The ancient Greeks recognized 48 constellations, including most of the familiar ones, and 40 more were added during the Age of Discovery, notably in the southern hemisphere, but with many filling in gaps in the northern sky. Though newcomers may find the names a bit of a mouthful, and are often amused that a collection of stars is meant to represent a beautiful princess or a mythological animal, the constellations have stood the test of time and

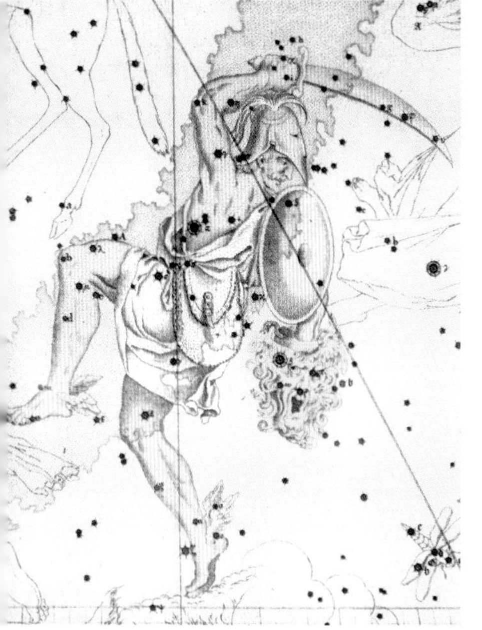

◀ *The mythological figure of Perseus as he appears in the atlas by John Bevis of about 1750. The stars only loosely fit the hero's shape.*

are much easier to use than the sky's coordinate system. There is a grid system in the sky, known as right ascension and declination. These are basically the same as longitude and latitude, but the binocular observer rarely needs to refer to them, so they will not be mentioned again in this book. The only time they may be needed is when giving the location of temporary objects such as comets. But these days, finder charts for such events are usually available online.

One measurement that you will need to understand is the magnitude system for describing star brightnesses. Again, this has been bequeathed to us from the ancient Greeks, who referred to the brightest and most splendid stars as being of the first magnitude, while the faintest were sixth magnitude. The system is therefore like the winners of prizes, with the top performer getting the lowest rating in actual numbers, so a star of magnitude 5.5 is fainter than one of magnitude 3.7. The scale now extends way down through the numbers, with the very faintest objects detectable being around thirtieth magnitude. The very brightest objects have negative magnitudes. The brightest star, Sirius, is magnitude −1.6, while Venus can be as bright as magnitude −4.6. The stars in the Big Dipper are mostly magnitude 2, while from urban areas it is hard to see stars fainter than about magnitude 4.5.

The names given to stars merit a book in themselves, but for our purposes all you need to know is that in general the brightest stars in a constellation are given the Greek letters of the alphabet, Alpha, Beta, Gamma, and so on, usually in general order of brightness. Star maps often use the Greek characters for these, but here we are spelling out the English version of the names. The constellation name is then used in the Latin genitive version, so instead of, for example, Alpha Centaurus we have Alpha Centauri. In addition, many of the brighter stars have proper names, often derived from Arabic because it was Arabs who kept our knowledge of Greek astronomy alive during the Dark Ages. As a result, many of them have names beginning with "Al-," Arabic for "The."

A list of constellations and the meanings of the names is given on page 200.

Computer maps

The maps we have used are taken from the excellent free software program Stellarium, which is available for PC, Mac, or Linux. This gives a realistic view of the sky, including the Moon and planets, from any location, and is easy to use once you have got the hang of it. You can even change the appearance of the horizon to suit your location. Go to www.stellarium.org to download it (20 Mb).

January northern hemisphere
Looking south

January southern hemisphere
Looking north

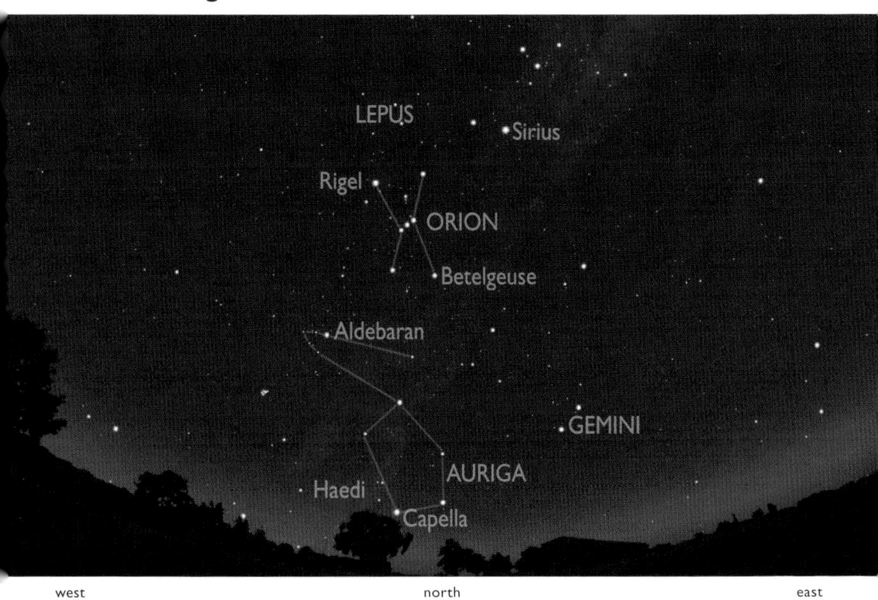

Looking north

Looking south

February northern hemisphere

Looking south

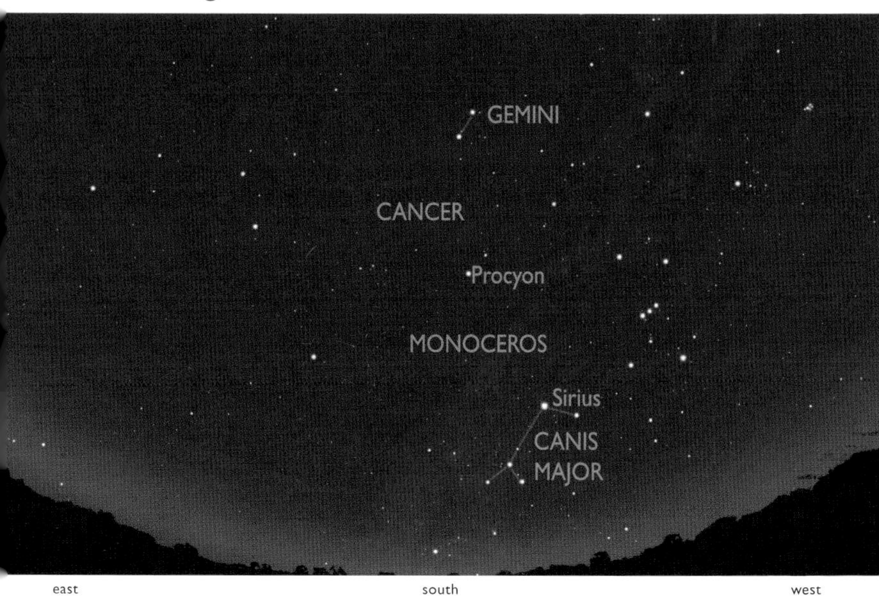

east south west

February southern hemisphere

Looking north

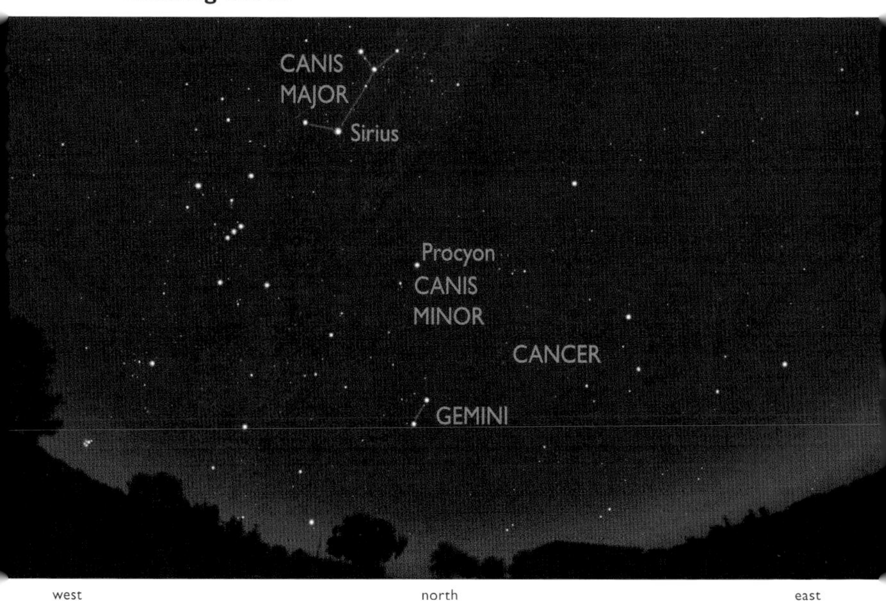

west north east

Looking north

Looking south

March northern hemisphere
Looking south

March southern hemisphere
Looking north

Looking north

Looking south

April northern hemisphere
Looking south

April southern hemisphere
Looking north

Looking north

Looking south

May northern hemisphere

Looking south

east south west

May southern hemisphere

Looking north

west north east

Looking north

west north east

Looking south

east south west

June northern hemisphere
Looking south

east south west

June southern hemisphere
Looking north

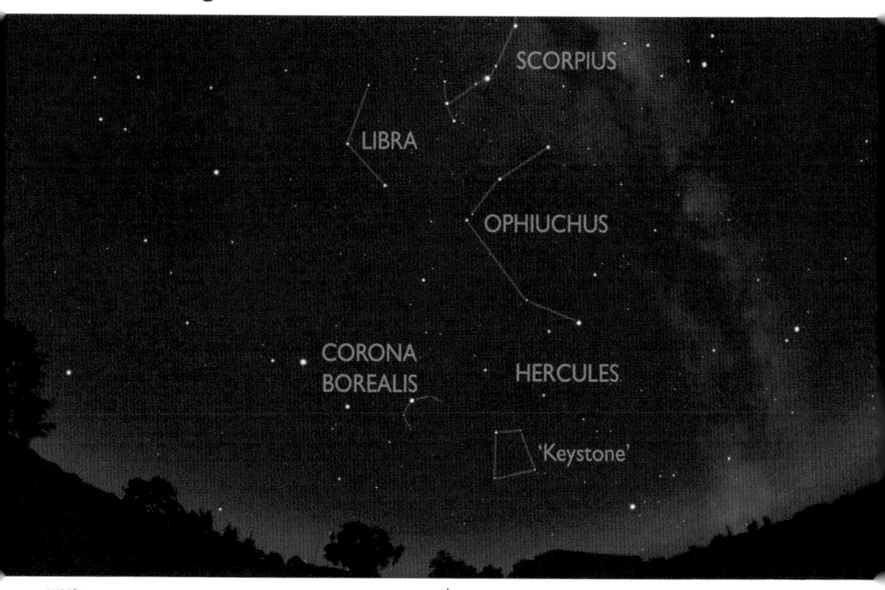

west north east

Looking north

west north east

Looking south

east south west

July northern hemisphere

Looking south

July southern hemisphere

Looking north

Looking north

Looking south

August northern hemisphere
Looking south

August southern hemisphere
Looking north

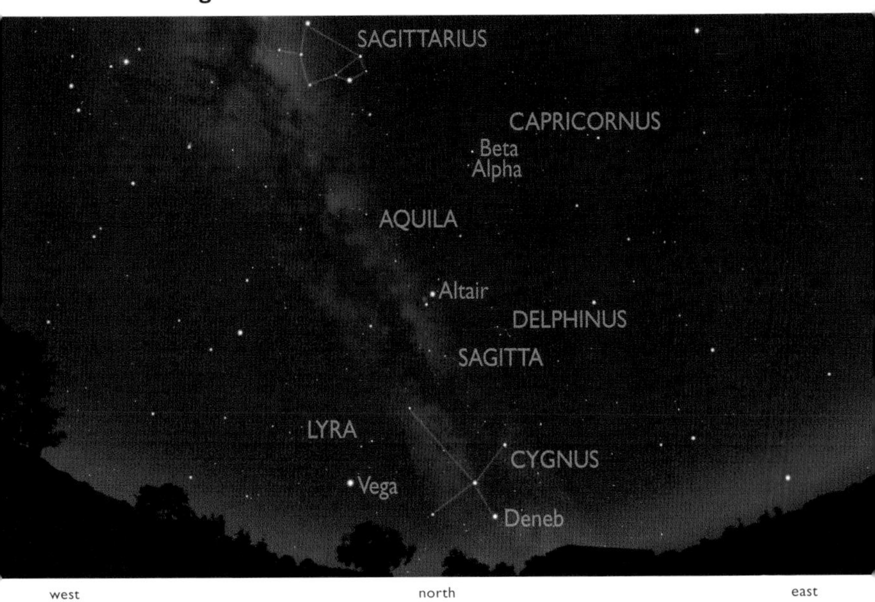

Looking north

Looking south

September northern hemisphere

Looking south

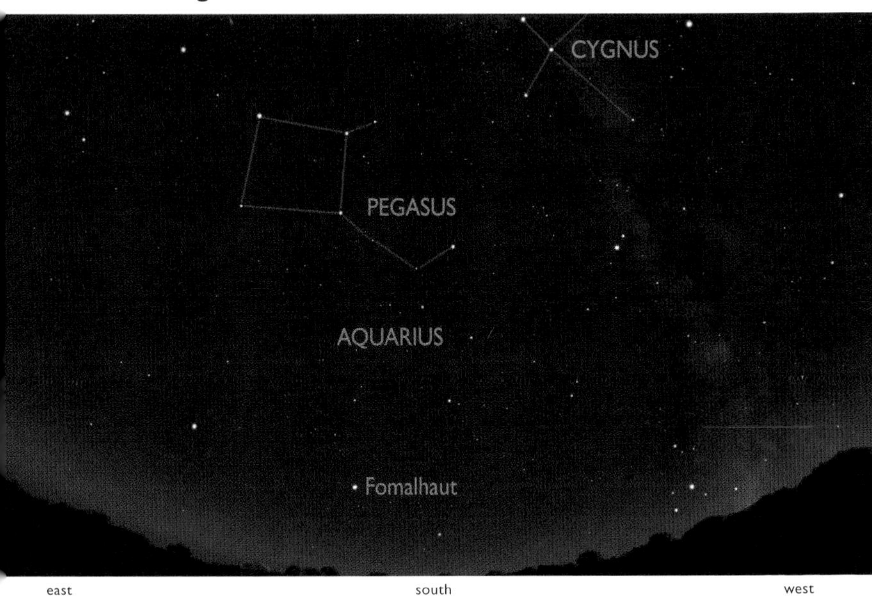

east south west

September southern hemisphere

Looking north

west north east

Looking north

Looking south

October northern hemisphere
Looking south

October southern hemisphere
Looking north

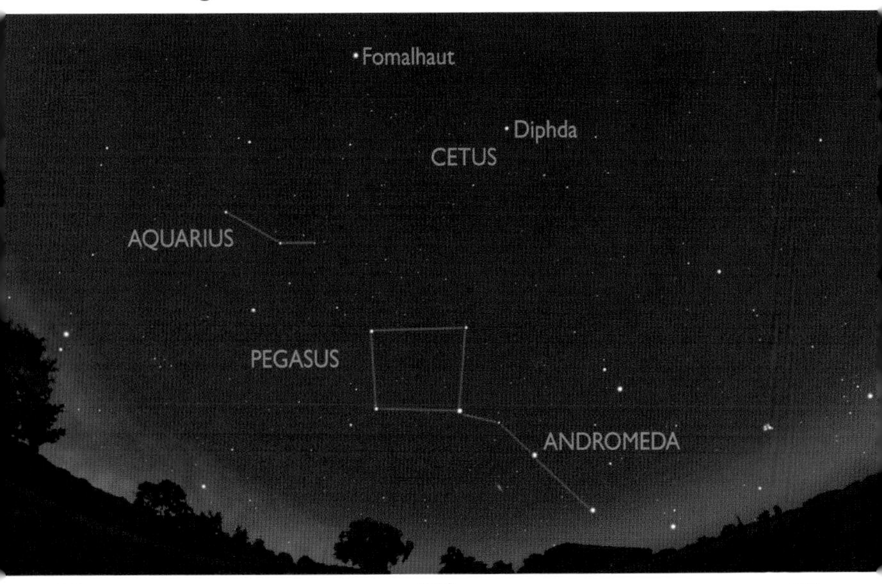

Looking north

west north east

Looking south

east south west

November northern hemisphere

Looking south

November southern hemisphere

Looking north

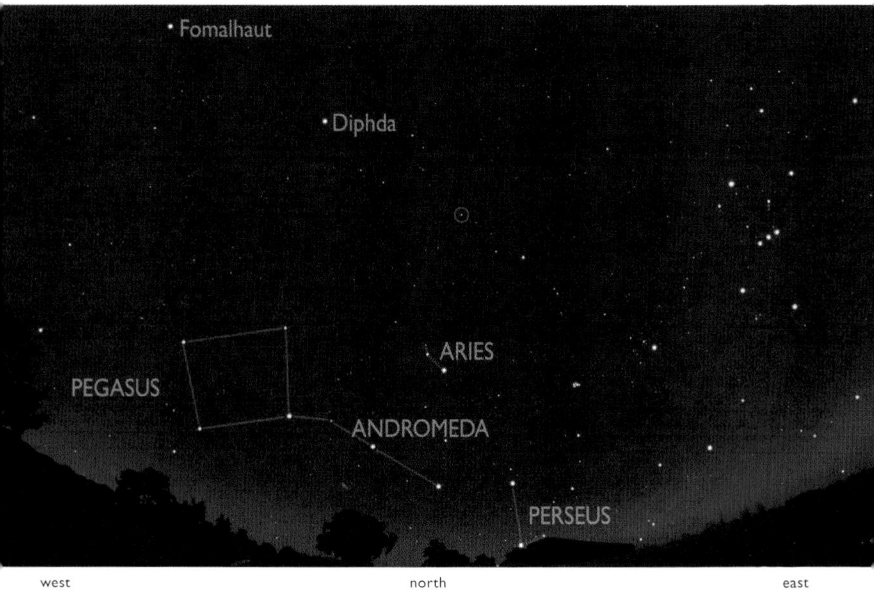

Looking north

west north east

Looking south

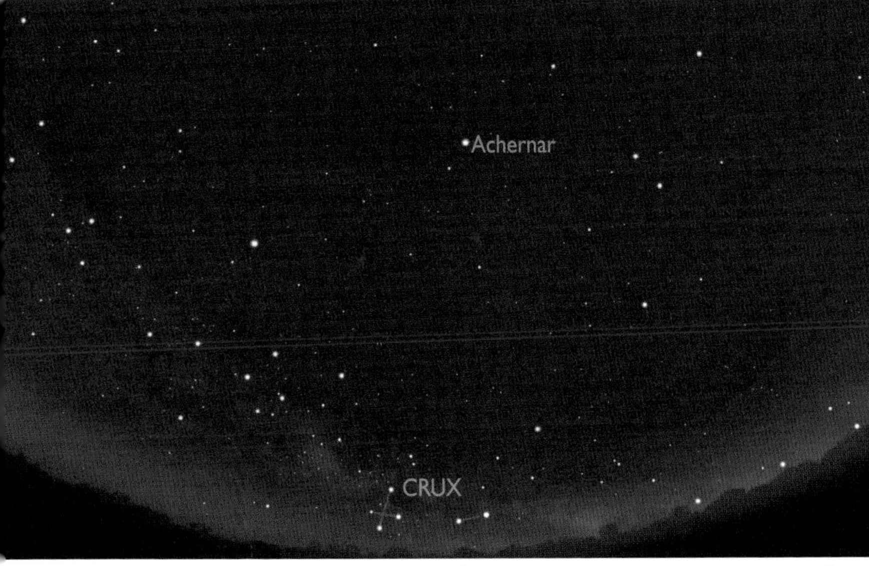

east south west

December northern hemisphere

Looking south

December southern hemisphere

Looking north

Looking north

Looking south

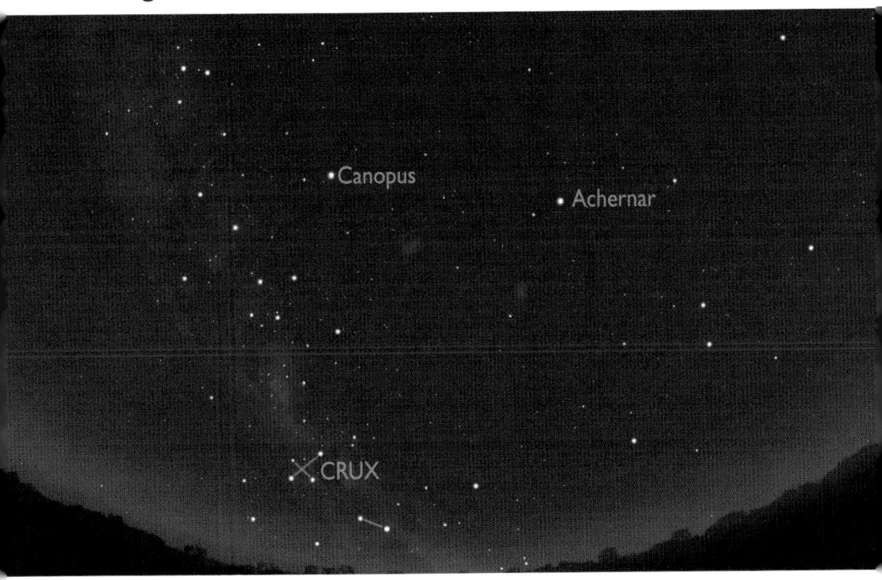

3 · THE BINOCULAR OBSERVER'S YEAR

The January sky

Orion is without a doubt the best signpost in the sky, and it is riding high in January. Even if you have never made any sense of a star chart, you can't fail to find Orion, which is the brightest of all constellations in terms of the number of bright stars it contains. With three stars of almost equal brightness equally spaced along a line, flanked by four others, it is instantly recognizable. It straddles the sky's equator, so it is visible from every part of the world.

To the people of the northern hemisphere Orion is associated with the depths of winter, and its stars glitter on frosty nights. Most of them are noticeably blue in color, which is in keeping with the chill in the air. But in the southern hemisphere Orion is a summer sight, and the brilliance of the blue-hot stars seems to add to the heat of the nights. In reality, the light from all the stars makes no measurable contribution to the warmth of our planet.

The line of three stars is known as Orion's Belt, which refers to its position on the mythological figure of the hunter striding through the sky. By good fortune, the stars point northwest to another bright star, the reddish Aldebaran in Taurus, and southeast to the brilliant Sirius, the brightest star in the night sky, in Canis Major. Surrounding Orion are Auriga the Charioteer to the north, Gemini the Twins to the northeast, and Lepus the Hare to the south below his feet. The mythological twins of Gemini have the same names—Castor and Pollux—as the two stars that mark their heads.

All of these can be spotted just by gazing. Orion is as much of a spectacle in its own right as anything you can see using binoculars or a telescope. Every astronomer, even the most experienced, will stop and stare whenever Orion is on the meridian. Orion is amazingly symmetrical, and is far more than just a random collection of stars. Just how much this symmetry influenced our forefathers is a matter for discussion. There have even been suggestions that the positions of the Great Pyramids reflect on the ground the arrangement of Orion's stars in the sky, though few experts would agree with this. Nevertheless, it illustrates how strong is the appeal of this constellation. Observers south of about latitude 35°N can see a further bright star, Canopus, more or less directly south of Sirius but much farther south. Canopus is the second brightest star in the sky. And farther south still, just 30° from the sky's south pole, lies one of the greatest sights in the whole sky—the Large Magellanic Cloud, nothing less than a galaxy in its own right. You really need to be south of the equator to see this properly, as it is only about as bright as the Milky Way.

M42—the Great Orion Nebula

Your tour of the January sky has to start with what is definitely the best-known deep sky object of the lot—the Orion Nebula. This is one of the few nebulae visible to the unaided eye, and in a really dark sky you can't help but notice that there is something a little different about the region to the south of Orion's Belt. In the middle of the area bounded by the Belt stars and the two southerly flanking stars, Rigel and Saiph, lies a line of fairly faint stars which positively seem to glow. The eye alone can't really make out what is happening here, but this is where binoculars come into their own. You can now see that the central region of the line of stars consists of an extended mistiness. This is the Orion Nebula, which is one of the key objects of interest to astronomers both amateur and professional.

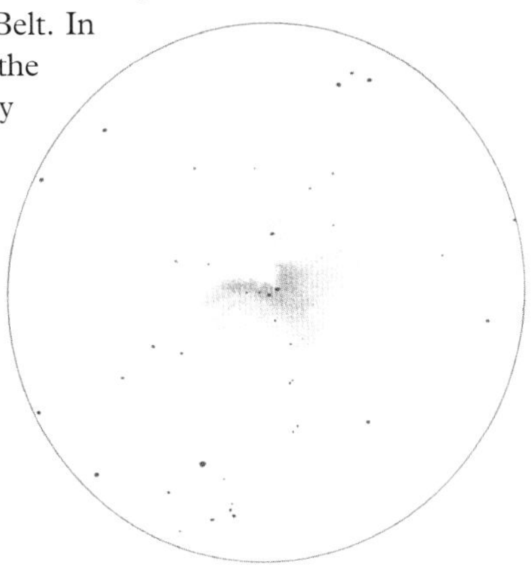

▲ *The Orion Nebula, M42, as seen through 15 × 70 binoculars by Michael Hezzlewood.*

The Orion Nebula is just the brightest part of a huge cloud of

▲ *In photographs, the Orion Nebula has a reddish color that can't usually be seen. The field of view here is 5° from top to bottom.*

gas that covers virtually all of the constellation of Orion. You can imagine the Orion Nebula as the tip of an iceberg—the only visible part of a much larger whole that lies hidden. Gas in space needs something to illuminate it before it will shine. The Orion Nebula glows because of newly born stars in its vicinity, and new stars are currently being hatched within it. However, the timescale is too long for us to be able to witness new stars bursting forth.

In binoculars, the view from urban areas is admittedly disappointing, though in some cases it is amazing that you can make out anything at all. But the nebula is bright enough to cut through all but the worst light pollution. At its center is a tight group of four stars known as the Trapezium, too close together to be seen separately other than in larger astronomical binoculars, such as 20 × 60s, though in smaller instruments the group looks distinctly fatter than the other stars. In darker skies using 10 × 50s it is possible to see the extensions on each side of the nebula's center: it is these features that make the nebula so distinctive. There is also a hint of the convoluted detail that makes this object such a magnet for viewing with a telescope.

In truly dark skies with a 20 × 60 or larger instrument the view is very detailed and you may be able to see the smaller and fainter M43 to its north. To find out more about "M" numbers, see page 58.

M35—a star cluster in Gemini

Although M35 is not the brightest cluster in the sky, it is easier to see than many others in binoculars because of its size, about the diameter of the full Moon. Find it by looking for the pair of stars, Mu and Eta Geminorum, that mark the end of the ragged line of stars extending

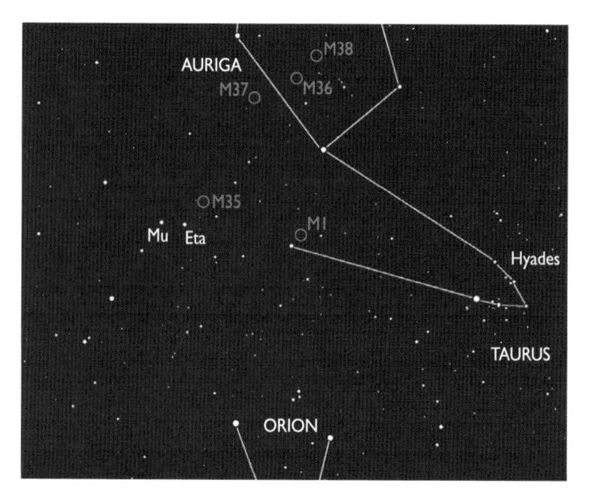

◀ *The skies to the immediate north of Orion include several objects visible with binoculars. The V shape of the Hyades cluster acts as a useful guide.*

from the star Castor and which represent the left foot of the Heavenly Twin of the same name. The cluster forms a flat triangle with these two stars. Because it is so large, you can pick out individual stars using binoculars, though to the naked eye it appears as a hazy patch in dark skies. Look carefully and you can see festoons of stars within the cluster. It lies about 2800 light years away and contains about 200 stars. A couple of the brighter ones are reddish in color, though one of these is an eighth-magnitude foreground star.

Nearby Eta Geminorum is also reddish in color and is a red giant star that varies in brightness between magnitudes 3.1 and 3.9 over seven months.

M1—the elusive Crab

The Crab Nebula has, in the jargon of today, good PR. It is almost as if some wily public relations executive for Nebulacorp has said "Let's try and sell this object. Hardly anyone can see it, but we'll make them all mad for it." Maybe it's the catchy name, maybe it's the interesting history, and probably it's helped by the dramatic photographs of the great stellar explosion, but people are keen to see it. The good news is that it is visible in binoculars—just—but you may need a magnification more than about 12× or so, and you need fairly good skies. Some people can see it with 7 × 50s or smaller binoculars, but many of us will be hard pressed to see anything at all. Even telescope users struggle to find it if they live in a moderately light polluted area. Faith Jordan, using 8 × 42 Leica binoculars from a dark country site in the UK, describes it as "Faintly visible as a barely discernible smudge."

▼ A finder chart for the Crab Nebula, centered on Zeta Tauri. The field of view has a diameter of 5°.

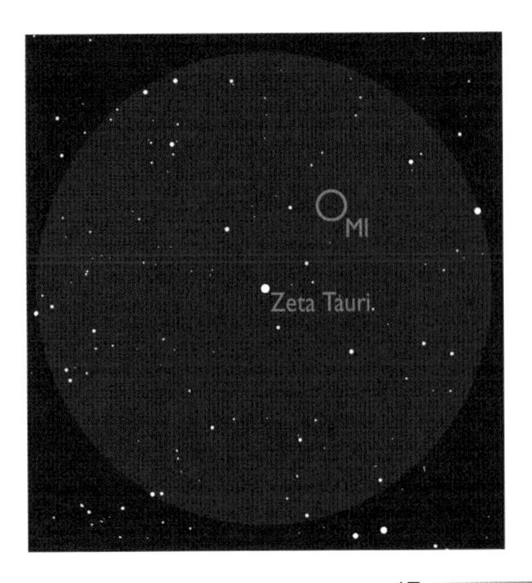

Fortunately, the Crab Nebula is easy to locate. It is a faint and tiny smudge just over 1° to the northwest of Zeta Tauri, which is the lower of the two horns of Taurus.

At this position, on July 4, 1054, a brilliant object appeared in the eastern sky just before sunrise. Rather than being the planet Venus, which was then in the evening sky, it was actually a supernova—a giant star that had collapsed on itself, resulting in a sudden brilliant explosion. The remains are still visible as the Crab Nebula, a name which was bestowed on it by Lord Rosse when he examined it with a giant telescope in the 1840s. The filaments and streamers which are so spectacular in photos are not visible in binoculars, but we can at least see the remains with a bit of luck. To astrophysicists it is a treasure trove of information about what happens when a massive star reaches the end of its normal life.

M36, M37, and M38—triple delights

North of Orion lies Auriga, a pentagon of stars—though the most southerly of these is shared with Taurus as it marks the tip of the more northerly horn of the Bull. Scan the area with your binoculars and you will soon spot a trio of small and distant star clusters—M36, M37, and M38. Actually, M36 is the central of the three. With binoculars these clusters appear as fuzzballs rather than as groupings of stars. They are all around 4000 light years away, and each one is subtly different. M36 is the smallest but brightest, while M38 is largest but most difficult to see in poor skies. You can just pick them out with the naked eye in good skies.

Capella and the Haedi

The bright star Capella, at the top of the pentagon of Auriga, is curiously referred to as the Goat Star, as Auriga is shown on the mythological figures as carrying a she-goat and her kids—represented by the

▲ *Capella and the Haedi. Capella is a yellow giant star, and its color is noticeable compared with the blue stars of Orion.*

triangle of stars just to the west of Auriga. In fact only the more southerly two are genuine Haedi, though many people refer to all three as such. They are worth a look with binoculars as they form a pretty group. The most northerly of the three, Epsilon Aurigae, is a noted variable star with a mystery attached to it. Every 27 years it undergoes a dimming in brightness lasting nearly two years, such as between August 6, 2009, and May 15, 2011. These dimmings are undoubtedly caused by another body in orbit around Epsilon, but its exact nature is in doubt. It may be another star with a disc of material surrounding it, from which planets will eventually form, but only more careful observations by the world's observatories will help to solve the mystery.

Sirius and two clusters

The constellation of Canis Major is noted for Sirius, the brightest star in the night sky (notice the qualification "night," the Sun is the brightest star in the sky!). Though all stars appear as points of light, Sirius is a star that appears to have a life of its own by virtue of its brilliance. When it is low in the sky in particular, as is often the case from the northern hemisphere, it appears to be constantly in motion, sending out vivid colored sparks and rays. This is nothing more than twinkling, which is caused by the path of its light being distorted by our turbulent atmosphere, but in the case of Sirius we see the effect more strongly than with any other star. If you can stand the glare, it is fascinating to watch its contortions for a while with binoculars. Occasionally people believe that they have seen a UFO when Sirius is particularly active.

Sirius is referred to as the Dog Star, from its position in Canis Major, the greater of the two dogs that follow Orion, the Hunter, through the sky. In ancient Greek times, it was thought that in July and August the heat from Sirius in the daytime sky combined with that of the Sun, giving the hottest days of northern summer. We still refer to the hottest days of summer as Dog Days for this reason. Of course, there is no perceptible heat from Sirius received on Earth.

Look just 4° south of Sirius—usually less than one field of view—and you come to the cluster M41. This is visible to the naked eye under the right conditions, and is therefore easy to spot with binoculars. Like M35, it is quite large and individual stars are easy to make out, though it is not as rich in faint stars as M35. Again, there is a red star near the center of the cluster, this time probably a genuine member. The Greek philosopher Aristotle seems to have referred to this cluster—he comments on a star with a tail "in the thigh of the Dog" which "If you fixed your sight on it its light was dim, but if you just glanced at it, it

▲ *Sirius (top) and the open cluster M41, 4° to its south.*

appeared brighter." This is a trick well known to astronomers, who refer to it as *averted vision*. Try it yourself on M41 or indeed on any faint object. Look directly at it and you can only see so much—or, in the case of faint nebulae, nothing at all. But look slightly away from it, and it appears to get brighter. This is because of the arrangement of light-sensitive cells in the eye. Your center of vision gives the most detail, but away from this your vision is actually more sensitive to light though you can see less detail.

Having found M41, if your skies are good you can now try for a fainter cluster in the adjacent constellation of Monoceros. If Sirius is the center of a clock face, M41 is at 6 o'clock as seen from the northern hemisphere when it is on the meridian. At 11 o'clock is a fourth-magnitude star, Theta Canis Majoris. Move toward that and then extend the movement the same distance again. The open cluster M50 should now be close to the center of your field of view. It is about one quarter of the size of M41 and has no star brighter than eighth magnitude, so you may see it as just a faint haze with most binoculars.

Southern hemisphere only

Large Magellanic Cloud and the Tarantula Nebula

The Large Magellanic Cloud, often abbreviated to LMC, is often described as being like a piece of Milky Way that has broken off. This is an apt description, though it is a Milky Way in its own right—a separate galaxy, some 170,000 light years away and a smaller companion of our own Milky Way galaxy.

It is an elliptical glow in the sky, much larger than any other deep sky object at about 6° in diameter, so it fills the field of view of most binoculars. But no stars are visible within this area—just subtle mottlings that large telescopes reveal to be mostly star clusters. But the brightest of

What is a nebula?

The word "nebula" is simply Latin for cloud. Nebulae (the plural, pronounced "neb-you-lee") are clouds of mostly hydrogen gas in space. They can be either bright or dark depending on whether or not they are illuminated by a star. Bright nebulae generally appear pink in photographs, because this is the color of glowing hydrogen, but you can also get reflection nebulae which appear blue in photographs. These contain a lot of dust given off by previous generations of stars, and reflect blue light for exactly the same reason that wood smoke, or indeed Earth's atmosphere, appears blue—the scattering of light. Gases other than hydrogen, notably oxygen and nitrogen, appear greenish in photographs. Visually, so-called bright nebulae are actually too faint, even when seen through a large telescope, to show color, though to some people they may appear greenish.

We can only see the dark nebulae against something brighter—either a bright nebula, or a background of stars. The Orion Nebula on this page is a good example of a starbirth nebula, a cloud of gas dense enough for new stars to be forming from it. The other main type of nebula is a planetary nebula (see page 80).

▶ The Trifid Nebula, M20, photographed by Philip Perkins. This shows the three types of nebula—a pink hydrogen gas or emission nebula, a blue reflection nebula, and dark nebulae which here form lanes and give rise to the nebula's name, meaning "cleft in three."

them is not a cluster, and is known as the Tarantula Nebula. This is a starbirth region that puts the Orion Nebula into a very minor league. It is often said that if we were at the same distance from the Tarantula as we are from the Orion Nebula, it would cover 30° of sky and would be bright enough to cast shadows. In binoculars it appears just as a brighter elliptical patch. Only in telescopes and photographs do the extensions that give it the appearance of a tarantula spider become obvious.

The LMC and its smaller neighbor, the Small Magellanic Cloud (see page 106) are in orbit around our Galaxy and will in the future be drawn into it by gravity. This is a fate that awaits many small galaxies close to larger ones.

The February Sky

February's skies are less glittering than those of January. Orion is moving toward the west, though the twin stars of Gemini are prominent in the northern sky, with Procyon to their south, near the celestial equator. The stars of Canis Major are farther south still, and moving well into the southern hemisphere are some of the scattered stars of Vela and Carina. There are no good signposts in this part of the sky other than those for January, and much of the area is filled with rather faint constellations such as Monoceros and Cancer.

Despite being a constellation of the zodiac, Cancer is very insignificant. The twelve signs of the zodiac are well known as astrological signs, but that is another matter. At one time, the stars were believed to affect the lives of people on Earth, and the path of the planets through the sky, the ecliptic, was divided up arbitrarily into twelve. So the constellations that lie along the ecliptic have received undue prominence. Cancer is an example of a faint group of stars that has become famous by chance. None of its stars are bright, and though it does contain one of the binocular showpieces of the sky, this has nothing to do with its fame.

▼ *In February skies, use Gemini and Leo as signposts. The Head of Hydra is the pattern of stars at the center of the map, below M67.*

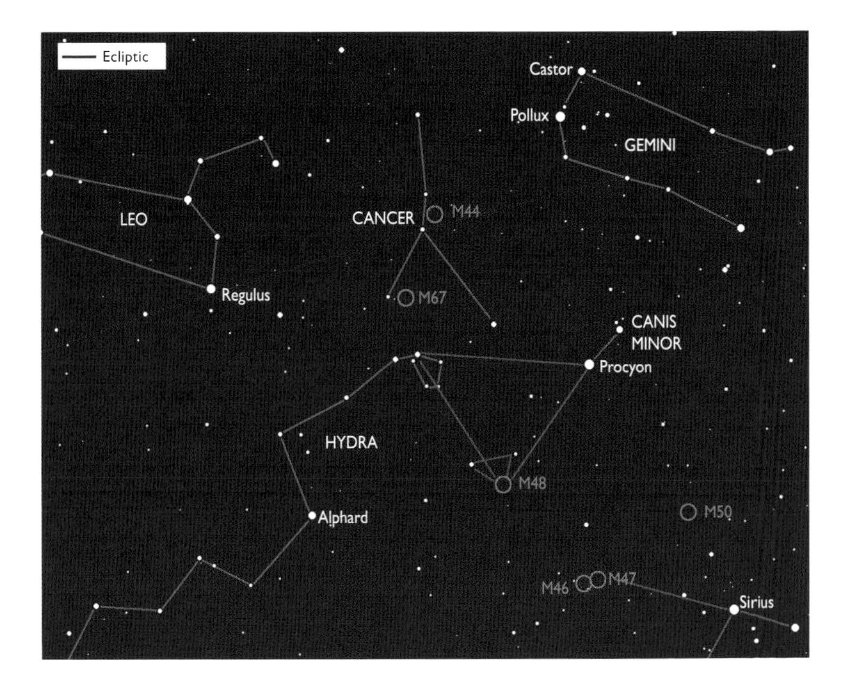

The Milky Way in the February sky is fainter than anywhere else along its length. It is hardly noticeable, compared with its appearance in July and August, though it starts to brighten up as you travel farther south. And to the north of Gemini, the sky is particularly barren.

M44—the Beehive cluster

Even a bright Moon or light pollution is not enough to drown out the cluster M44. It is a fine sight in binoculars of any size, which are indeed the best way to see it on account of its size—about 1° across. Its brightest stars are of sixth magnitude, so even though individually they are too faint to be picked out with the naked eye, together they form a distinct fuzzy patch in the sky. In fact it was Galileo himself who was the first to see that it was a star cluster: prior to the invention of the telescope it was just another mystery.

Two fourth-magnitude stars flank the cluster to its east. In the past these two stars were known as the donkeys, as suggested by their names—Asellus Borealis and Asellus Australis, the northern and southern donkeys. (The words borealis and australis, Latin for north and south, often crop up in astronomical terminology.) As a result, the fuzzy glow, which was obvious to early astronomers, became known as the Manger, or Praesepe in Latin (usually pronounced "Prye-see-pee"). Its appearance also gave its more modern name of the Beehive, a good description. The cluster lies about 580 light years away.

With a dozen or so stars of sixth magnitude, it is among the brightest of all clusters. There are three wide pairs of stars which are worth looking for. If you fancy trying your hand at drawing a star cluster, this is an excellent one to start with, as the bright members are well scattered and form a good basis from which to fill in the fainter ones.

▼ *This photograph of Cancer shows both M44 and M67.*

M67—the ancient one

There is another star cluster in Cancer which is accessible with binoculars, M67. This is smaller and fainter than M44, and is also very ancient at about four

Binocular special—NGC 2451

There are not many objects that are suited only to binoculars rather than telescopes, but NGC 2451 in Puppis is one of them. This object is roughly midway between a pair of stars, Pi and Zeta Puppis, which are best found from the much more familiar pattern of Canis Major to the north. Puppis is a constellation that sounds as if it should have something to do with dogs as well, but in fact the name refers to the poop deck or stern of a ship, Argo Navis, that was the mythological vessel of Jason and the Argonauts. It was a huge, sprawling pattern, and is now split into more manageable components, Puppis (the Poop), Carina (the Keel), and Vela (the Sails). Most of the rest of Argo Navis is below the horizon from northern Europe, and this is an object for those living south of about latitude 40°N.

The reason why NGC 2451 is so suited to binoculars is that it is a very sparse cluster, somewhat dim, though easily spotted with the naked eye, but too scattered to show anything worthwhile through a telescope. It has several members of fifth and sixth magnitude, which make it easy to spot, with several bright jewels to make up for its lack of faint stars. Look for the color difference between its red central star and the others.

billion years, making it one of the oldest in the Galaxy. It is also one of the largest in the Galaxy, containing as it does more than 500 stars in all, and is about five times more distant than M44. Find it by first locating Alpha Cancri, which is to the southeast of the two donkeys. There is a fainter star just to its west, then M67 is a greater distance to the west of that, all within a binocular field of view. There is a magnitude-7.8 star just to the east of M67, but it is not a member of the cluster. Some people can spot individual stars in M67 using binoculars, but to others it remains a haze. The brightest members are about tenth magnitude.

The Head of Hydra and M48

The cluster M48 is not far from those in Cancer, but is less well indicated by stars. To find it, look to the south of Cancer for a pattern of six stars of about fourth magnitude, looking like a very deep saucepan. This is known as the Head of Hydra, Hydra being a long snaking ribbon of stars that covers an enormous area of sky, as befits the water monster that it represents. Oddly enough, although the mythological beast had nine heads, the one in the sky has only the one. But it is an attractive group in an otherwise barren area of sky. The only other bright feature of Hydra is the star Alphard, to the southeast of the head, which is second magnitude. The name Alphard translates from Arabic as "the Solitary One," which is a comment on its location far from any other bright stars. Hydra is still with us in June, such is the extent of this constellation.

Having found the Head of Hydra, make a south-pointing equilateral triangle with Procyon as the other corner, and you should spot M48. It, in turn, is at the base of a smaller triangle with a pair of fourth-magnitude stars, the eastern of which is flanked by two fainter stars. It is much easier to make out individual stars in M48 compared with M67, and this is an object for which you will benefit from using higher-magnification binoculars.

M47 and M46—a contrasting pair

This adjacent pair of clusters is on a line taken from Beta Canis Majoris through Sirius, extended about two and one half times the distance between them. M47 is by far the easier of the two, and is visible with the naked eye under good conditions, though binoculars are needed to see it in average skies for many people. It has a scattering of fifth- to seventh-magnitude stars, plus some fainter ones, so it is a sight that looks increasingly better the darker the sky. The adjacent M46 consists mostly of tenth-magnitude stars, and is a challenge unless you have dark skies.

▲ *Clusters M46 (left) and M47: a photograph that shows stars down to thirteenth magnitude and covers 2° of sky from top to bottom.*

Southern hemisphere only

IC 2391—the Omicron Velorum cluster

Some people describe this as an asterism—a collection of stars—rather than a cluster, but it is a genuine grouping of stars that are all at roughly the same distance of about 500 light years. It was first recorded as long ago as the tenth century by the Persian astronomer Al-Sufi, along with some better-known objects such as the Andromeda Galaxy and the Large Magellanic Cloud. He referred to the cluster as a "nebulous star," which is how it appears to the naked eye. But binoculars show it as a group of a dozen or so stars brighter than ninth magnitude, the brightest of which is Omicron Velorum itself at magnitude 3.5.

The group is very pretty in binoculars, and you need no great effort to find it, even in suburban skies. It lies to one side of a feature known as the False Cross—a pattern of four stars (marked in blue on the map on page 54) in Vela and Carina lying about 30° to the west of the genuine Southern Cross (Crux). The False Cross does resemble the real one, but it is half as big again and points in a completely different direction. The Southern Cross acts as a useful guide to the location of the south celestial pole (see page 67), so it is worth learning the difference!

Gamma Velorum and NGC 2547

The star Gamma Velorum is the brightest in Vela, and sits by itself to the east of Canopus. The cross-arm of the False Cross points to it. To the eye it is just a bright star, but turn binoculars on it and you will see two stars, very close together. If you are using lower-power instruments, such as 6 × 30s, you may have difficulty seeing the two stars, particularly if either the glasses or your eyes are not of good quality, but with higher magnifications they should be easy, particularly if you steady the binoculars.

The brighter of the two is one of the hottest stars in the sky and is noticeably bluish. While the Sun is at a temperature of 5700°C (10,300°F), Gamma Velorum is a searing 60,000°C (108,000°F). It is of a rare type known as a Wolf-Rayet star, and is the brightest example of such a star in the sky. It has a companion, too close to be made out separately even with a telescope, which is also a massive and brilliant star. The companion which you can see with binoculars is over two magnitudes fainter and is probably more distant than the brighter star.

Gamma Velorum is often referred to as Regor. This name dates back to just the 1960s, when it was jokingly applied to the star by American astronaut Gus Grissom after fellow astronaut Roger Chaffee. Regor is Roger spelt backward. Both Grissom and Chaffee died in the Apollo 1 capsule fire, and the name is now used in honor of the dead astronauts.

Just 2° to the south of Gamma Velorum is another binocular-only cluster, NGC 2547. Its stars are mostly seventh and eighth magnitude, and cover an area of about ½°.

NGC 2516

The False Cross also leads you to one of the glittering jewels of the southern sky, NGC 2516. It lies some 3° from Epsilon Carinae, the star

► *A sketch of multiple star Gamma Velorum, as seen through 10 × 50 binoculars by Jeremy Perez of Flagstaff, Arizona.*

M, NGC, or IC?

The catalog numbers given to most deep-sky objects are usually either M numbers or NGC numbers. The M stands for Messier, an eighteenth-century comet hunter who made a list of the brightest objects that could be confused with comets, though his list does also contain some that would never be mistaken for comets, such as M45, the Pleiades. In general, Messier objects are the brightest deep sky objects and a good number can be seen with binoculars. A far more comprehensive list of the fainter objects was subsequently made by William Herschel. This was then published in an augmented form by J L E Dreyer in the nineteenth century as the *New General Catalogue*, or NGC. Some subsequent additions are known as the *Index Catalogue*, or IC.

at the bottom of the long axis of the cross, and is a naked-eye object in even moderately good skies. With low-power binoculars it appears as a haze with one or two fifth-magnitude stars visible; use 10× or more and additional stars appear. You can spend some time trying to pick out fainter and fainter stars.

The March sky

Leo the Lion is the dominant constellation in the March sky. Its shape really does resemble a crouching lion, certainly as seen from the northern hemisphere. The bright star Regulus (Alpha Leonis) marks the lion's heart, with a curve of stars indicating the line of its mane. The pattern seems to have reminded people of a lion for thousands of years, as even the earliest Mesopotamian constellations include the figure.

The ecliptic runs through Leo, so it is possible that one of the bright planets may be seen within its borders. Regulus itself is one of the few bright stars in front of which the Moon can pass—an event known as an occultation. There may be several of these in any one year, but each one is visible from only a limited area of the Earth's surface.

This area of sky is particularly barren of binocular objects except in its most southerly part, where there is a section of the Milky Way. However, two of the few galaxies visible with binoculars—M81 and M82—are high in northern March skies.

Northern hemisphere only

M81 and M82—Galaxy twins

Where galaxies are concerned, it helps to wait until they are at their highest in the sky if you want to see them with binoculars, particularly if there is any light pollution around. Fortunately, this pair of galaxies is very high in the March skies as seen from Europe and North America. They are well worth the search, and are comparatively easy to spot once you know where to look, even though they are not particularly well signposted. While not large, these galaxies are of a fair size compared with most that you can see with binoculars. Each is about the size of a typical sea on the Moon—the actual size you see depending on how good your skies are. And because there are two of them the appearance of one reinforces the other. One tiny fuzzy spot can get lost, but two attract the eye.

Your quest begins with the familiar shape of what is called the Big Dipper in North America and the Plow in the UK, but which, to modern eyes looks more like a saucepan. This features more prominently in the April sky section, but at this time of year it is high in northern hemisphere skies. Draw a diagonal between the bottom left star and the top right star of the bowl of the saucepan and continue the line the same distance again. You are now very close to a pair of stars, the brighter of which is the more distant from the bowl of the saucepan and is magnitude 4.5. However, if your diagonal is slightly off, you will come to another similar pair, slightly farther west. The wrong pair have another

▲ *To find M81 and M82, take a diagonal through the bowl of the saucepan as shown. The galaxies lie 10° farther along the line.*

◀ *A sketch of M81 and M82 as seen using 12 × 45 binoculars from Flackwell Heath, UK, by Robin Scagell.*

▼ *A photograph of the two galaxies taken from Kitt Peak Observatory, Arizona, shows the difference between a sketch and reality.*

star nearby, making a long triangle with them, marked on the map on page 59 with an X.

Depending on the field of view of your binoculars, if you place the correct pair of stars to the far side of the field of view from the saucepan, M81 and M82 will be near the middle of your field of view. They are unmistakable once you see them, because one of them—M82—is long and thin. One of its nicknames is "the Cigar Galaxy," which describes its shape very well. M81 has a brighter core, and is easier to see in low-power instruments. It is typical of galaxies that they are bright in the center and fade away to nothing at the edges. The darker your skies, the larger your binoculars, and the better your eyes, the more you will see of it.

These two galaxies are a genuine pair, the principal members of a small group of galaxies some 12 million light years away from us. It is tempting, when looking at galaxies, to speculate on the inevitability that there are beings within them who are looking out, perhaps using their own equivalent of binoculars, at our own Milky Way, which is of the same sort of size as M81 and therefore will look quite similar to them as M81 does to us. The two galaxies are considerably closer together than are our own Milky Way Galaxy and the Andromeda Galaxy, the two main members of our own Local Group. The denizens of M81 will see, in addition to their own milky way of stars, a great cigar shape of similar brightness, covering some 10° of their own sky, which is M82. The closeness of the two galaxies has led to gravitational interactions between them, which have created a shock wave of star formation within M82. This is very obvious in long-exposure photographs of the galaxy.

Southern hemisphere only

Eta Carinae Nebula and surrounding clusters

The Eta Carinae Nebula is the largest and brightest of all the nebulae, beating the Orion Nebula hands down in a straight fight. Due to its being easily seen only from the southern hemisphere it has a lower standing than it might. The star after which it is named is currently fairly insignificant, though that has been known to change.

Every southern-hemisphere observer recognizes the Southern Cross (Crux) and the two nearby stars, Alpha and Beta Centauri, which are highest in the sky on May evenings but nevertheless prominent during March. Eta Carinae is exactly opposite the Southern Cross from Alpha and Beta. There is no particularly bright star nearby, though there are numerous fainter stars in the general area. It is usually visible with the naked eye, though from very light polluted areas binoculars will help to establish its location. In a good sky it is simply a stunning sight.

The nebula appears a good 1° across in dark skies, with several stars within its area. There is a prominent L shaped dark lane, and Eta Carinae itself is a fifth-magnitude star just at the junction of the L, within the brightest part of the nebula. This star drew attention to itself in 1843 with an outburst of brightness to magnitude –1—brighter than any star except Sirius. It subsequently faded, and during the first half of the twentieth century it was only about magnitude 8. It has subsequently brightened, reaching a brief peak at 4.7 in mid 2006. Exactly how bright it will be when you read this can best be decided by going out and looking, if you are suitably located.

Eta Carinae is a very massive star, one of the most massive that we know of. Estimates of its mass depend on who you ask, but it could be as high as 150 times that of the Sun—though there are grounds for believing that such an enormous star cannot exist. Most experts believe that it cannot fail to become a supernova in due course, resulting in a spectacular explosion that could make it

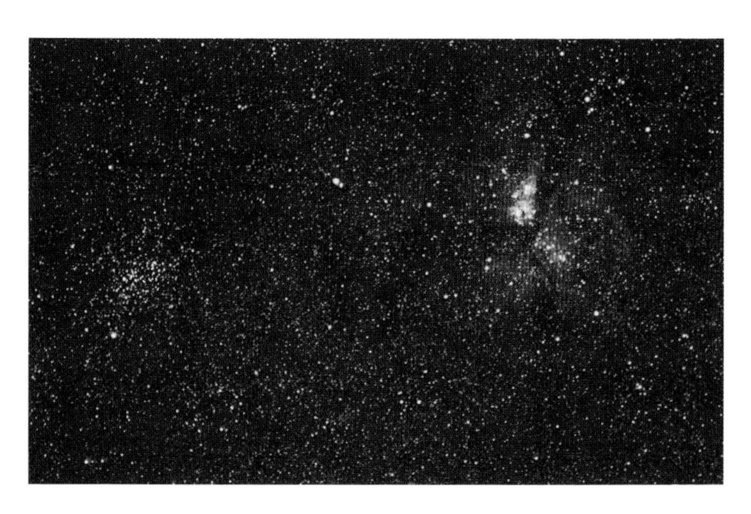

▲ *A sketch and photograph of Eta Carinae made by Chris Picking, of Wellington, New Zealand. The sketch was made using 10 × 50 binoculars. In the photograph, the cluster NGC 3532 can be seen to the left.*

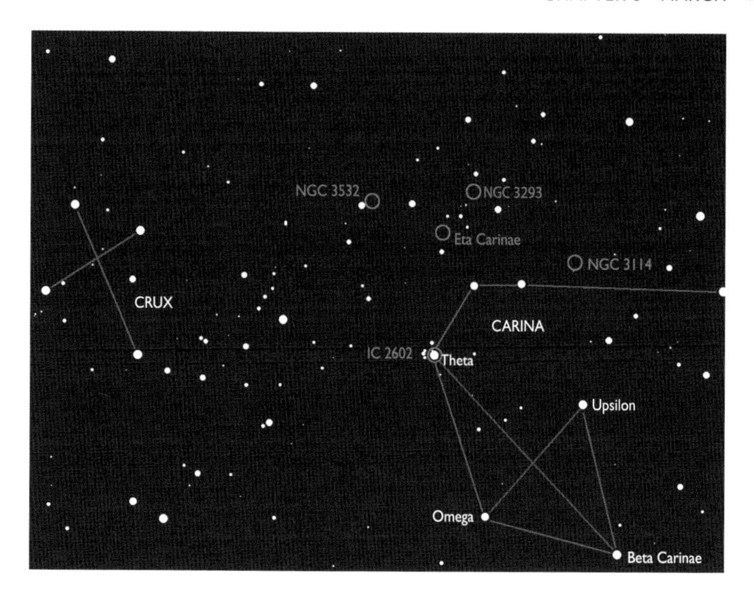

▲ *The Southern Cross (left) acts as a signpost for several bright binocular objects in the southern Milky Way. The Diamond Cross is shown lower right.*

the brightest object in the sky other than the Sun itself. Some speak of it becoming a hypernova, a super-exploding star whose brightness would temporarily exceed all the other stars in the Galaxy put together. If that happened, it would shower the Earth with gamma rays. Fortunately our atmosphere would protect life on the surface of the planet, but any astronauts in space or on the Moon at the time would be unprotected, even in a spacecraft. There would be little or no warning, either. So how long will it be before Eta Carinae blows its top? The best guess is "sometime within the next million years," though some venture that it could even be within the next 1000 years.

Like the Orion Nebula, the Eta Carinae Nebula is a starbirth region. There is a concentration of faint stars within its area and surrounding it. One cluster in particular is amongst the finest in the heavens, NGC 3532, just to its east. Though none of its stars is brighter than sixth magnitude, there is a wide scattering of seventh- and eighth-magnitude stars within its elliptical shape that make it a real binocular treat. It may not have the brilliance of the Pleiades, but the richness of this cluster makes it unforgettable. It is visible with the naked eye even in light polluted areas.

On the opposite side of the nebula lies the hardly less impressive NGC 3114, with several dozen stars that go down as far as your binoculars will show. It's easy to spot even in poor skies. And to the northwest of Eta Carinae, still within the same binocular field of view, lies a much tighter cluster, NGC 3293, which at low magnifications is an unresolved knot of stars. There are about 20 stars of between sixth and eighth magnitude within just five arc minutes. It is sometimes known as the Gem Cluster or the Diamonds and Ruby Cluster because of a strongly red star close to one edge.

The Diamond Cross and the Southern Pleiades

The southern sky is awash with crosses. In addition to the Southern Cross (Crux) and the False Cross, there is another, known as the Diamond Cross, which again has a similar shape, though as its name

About star clusters

As well as being attractive sights, star clusters are very useful to astrophysicists. In particular, they make it possible to see a uniform set of stars at a specific distance. Looking at the sky as a whole, we see scattered stars of all brightnesses and ages. It is not immediately obvious which are faint stars nearby, and which are bright stars at a great distance. However, stars in a cluster provide a fair certainty that all the stars are at a more or less uniform distance, so the differences between them are intrinsic to the stars rather than being due to different distances or maybe varying quantities of intervening material absorbing the light. There are inevitably one or two stars in the line of sight to the cluster that spoil the statistics, but they can usually be weeded out.

All the stars in a cluster will have formed from the same gas cloud, rather than forming separately and then gathering. They all have the same overall movement through space, and though they may have formed over a period of time, they are all essentially the same family of stars going through the same life cycles. The brightness of a star in the normal stages of its life depends on its mass, and in clusters we see a general trend from bright, massive stars that shine with a bluish light to less massive stars that are dimmer and are white or yellow. The dimmest stars are the least massive: they shine with a feeble red light.

In a young cluster of stars, such as the Pleiades in Taurus, there are plenty of bright stars shining furiously and using up their stock of hydrogen fuel. But bright stars die young, leaving the less spendthrift white, yellow, and red stars. M67 has no blue giant stars and is estimated to be about four billion years old, compared with only about 100 million years for the Pleiades. M44 in Cancer is older than the Pleiades, but still young, at about 750 million years.

These are all open or loose star clusters. Such clusters range from just a few stars that formed together to large groups of several hundred stars. The members of the smaller ones drift apart shortly after their formation and join the implies it is a perfect diamond. It is about the same size as the False Cross (see page 56) at 10° from end to end, and is similarly oriented at an angle to the Southern Cross, though its stars are fainter.

It consists of four stars in Carina, with the designations Beta, Omega, Theta, and Upsilon. Of these stars, Theta is a splendid object in binoculars. Surrounding third-magnitude Theta is a handful of fainter stars that have earned the group the name of the Southern Pleiades, from a general similarity to the brighter and better known cluster in Taurus (see page 113). Though you can see the individual stars if you have good eyesight, binoculars are needed to show them well, and will reveal some fainter stars, though not as many as in the genuine Pleiades. The cluster is also designated IC 2602.

general background; but the larger star clusters, such as M67, may remain held together by gravity. Open clusters are features of the spiral arms of the Galaxy.

There is another type of star cluster, known as a globular cluster. These consist of very much larger groups of stars, maybe up to a million or more, and have a spherical shape rather than being loose gatherings of stars. Instead of inhabiting the spiral arms like the open clusters they are to be found around the fringes of the Galaxy, like bees hovering around a hive. And unlike open clusters they consist of very old stars, with no hot, young, blue stars at all. The ages of stars in globular clusters are right at the other extreme from the open clusters—they are typically over 12 billion years old, which puts them among the oldest stars in the Universe, and almost as old as the Universe itself.

▲ *The Wild Duck Cluster (M11) in Scutum is a typical open cluster.*

▲ *Omega Centauri is the largest and most impressive globular cluster.*

The April sky

In both the northern and the southern hemispheres, April brings what are to their respective populations the best known constellations in the entire sky. They are also a great help for beginners in getting their bearings. Both are circumpolar for most of the populated areas of their hemispheres—that is, they are above the horizon all the time, and just change their orientation as the night goes by.

In the northern hemisphere lie the seven bright stars of Ursa Major, the Great Bear, better known as the Big Dipper in North America and as the Plouw in the UK. Both these implements are now antiquated, yet their names live on. As mentioned in the March description, the seven stars also strongly resemble a saucepan.

This saucepan constantly whirls round the sky's north pole, being at its lowest in the sky due north on October evenings, when it is most noticeable, and at its highest in April. By chance, the two easternmost stars of the bowl of the saucepan point almost exactly north, so they are referred to as the Pointers. By an even more happy chance, there is a bright star close to the sky's north pole—Polaris, the Pole Star. Over the centuries, a wobble of the Earth's axis will mean that Polaris will no longer be so close to the pole, but for the time being it is an invaluable guide to the north polar point and helps astronomers to get their bearings. Every beginner in the northern hemisphere should learn to pick out the Pole Star, which always remains at virtually the same point in their sky, and at the same altitude in degrees above the horizon as their latitude.

▲ *How to find the Pole Star using the bright stars of Ursa Major.*

On the celestial equator at this time of year, and therefore equally visible from both hemispheres, is the constellation of Virgo, the Virgin, with its bright star Spica. Traveling southward farther still we come to the quadrilateral known as Corvus, the Crow. Its stars are not particularly bright but as there are few other bright stars around, they stand out clearly.

Then, as far south of the equator as the Big Dipper is north, is the icon of the southern hemisphere—the Southern Cross (Crux). Smaller than many people might imagine, this compact

► *How to find the south celestial pole using the Southern Cross (Crux, right) and Pointers. The method is accurate to within about 3°—the circle marks the pole's true position. The dark Coal Sack Nebula is visible east of the Cross, with the Eta Carinae Nebula at far right, and IC 2602 nearby. This photo shows many more stars than are visible with the naked eye.*

little group is well known from its appearance on the flags of Australia, New Zealand, Brazil, and several other southern nations. It, too, has the useful knack of pointing toward the pole, though in this case there is no bright star to mark the spot. So anyone wishing to locate the south celestial pole must estimate four and one half Cross lengths to find the correct spot. A further check comes from the two stars to the east of the Cross, Alpha and Beta Centauri, which point to the Cross and are therefore sometimes also called the Pointers. But, unlike the northern Pointers, these two are at right angles to the pole, so bisecting them gives a line which crosses that indicated by the Southern Cross to mark the position of the pole itself—well, almost.

The naked eye is your best instrument to appreciate one of the deep sky objects in Crux—the Coal Sack Nebula. This is a dark nebula to the southeast of the Cross itself, and is seen against the background of the Milky Way, so you need fairly dark skies to be able to appreciate it.

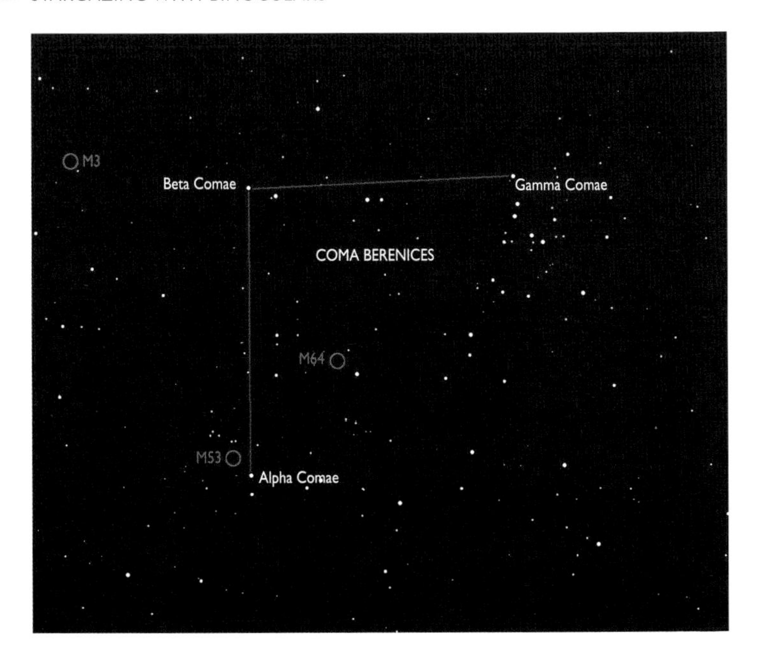

▲ *Binocular objects in Coma Berenices. The map also shows globular cluster M3, described on page 76. The Coma Star Cluster is below Gamma Comae.*

The Coma Star cluster—Melotte 111

One star cluster is so large and close that it is hard to contain it even within wide-field binoculars, and that is the Coma Star Cluster. This lies in the faint constellation of Coma Berenices, the Hair of Berenices, which has just three stars of fourth magnitude. But the Coma Star Cluster is a real treasure, with ten or so stars of naked-eye visibility within an area of about 4° of sky. Light pollution will wash it out to the naked-eye view, but with even moderately dark skies, this group stands out though you may not be able to pick out the individual stars. Binoculars show a mass of points of light filling the field, and the impression is rather better than you get through a telescope with many other clusters. The cluster is particularly close to us at about 270 light years. There is another Coma Cluster, this time of galaxies, but it is hard to see even with a telescope and is way beyond binocular reach.

Globular cluster M53

This is an easy globular cluster to locate, as it sits close to Alpha Comae Berenices, a fourth-magnitude star that is probably easiest found by looking midway between Denebola, the tail end of Leo, and Arcturus, the brilliant star in Boötes to the east. There is a sixth-magnitude star just south of Alpha Comae Berenices, and M53 is the same distance away but to the northeast. It appears as a little round fuzz, but no individual stars are visible in binoculars. As is the case with most globulars, a higher magnification helps. A sketch of M53 is shown on page 195.

A host of galaxies

April skies contain many distant galaxies that are on the borderline of binocular visibility for most observers. Anyone blessed with desert skies, large binoculars, and good eyesight could reel them off one by one. But the vast majority of observers would soon become frustrated with looking at yet another blank circle of sky, so we are going to limit the descriptions to just a handful of the brightest. Even these will be completely invisible to most observers in even moderately light polluted skies, but get the right night away from city lights and you could find that they suddenly leap into visibility.

Here is the Realm of the Galaxies—the Virgo Cluster. This is a huge supercluster of well over 1000 galaxies, of which the Local Group of galaxies that includes the Milky Way is a very outlying part. The cluster lies in a very convenient position more or less at right angles to our own Milky Way, so our view is unobstructed and as the area is barren of stars and gas clouds, the view is not spoiled, though it does make star hopping more tricky, as there are very few stars of even naked-eye visibility within the area.

You might imagine that with so many galaxies on view, you would stumble upon them with no difficulty, and indeed in perfect skies and with a suitable instrument this might well be the case. But the galaxies are distant—typically 55 or 60 million light years away, compared with 12 or 14 million light years for the easy objects such as M81 and M82, or the 2.5 million of the Andromeda Galaxy. This means that they are tiny in binoculars, typically only four or five arc minutes across, so the higher the magnification of your binoculars, the better.

Most of the galaxies display very little structure, even in a telescope, so don't hope to see spiral arms. It is enough to be able to spot these faint little fuzzy objects and realize that they are great galaxies, probably larger than the Milky Way, and contains billions of stars and planets, not to mention amazing sights about which we can only speculate. A fan of *Star Trek* might imagine that the *Enterprise* could visit these galaxies, but they would be wrong. The galaxies are so distant that even at Warp Nine, it would take many generations of crew before getting anywhere near the Virgo Cluster. And as far as the greater Universe is concerned that is just our neighborhood. But you can visit them on a sparklingly clear night away from city lights using powerful binoculars.

A tour of April galaxies

Start with a galaxy that is not actually a member of the Virgo Cluster as such, but is less than half the distance of the main members of that cluster at about 24 million light years. This is M64, also known as the Black Eye Galaxy from its appearance in telescopes. Having found Alpha Comae Berenices, move roughly halfway between that star and the Coma Star Cluster to find fifth-magnitude 35 Comae Berenices. M64 is at the same angle and distance from this star as M53 is from Alpha, though it is smaller and fainter. You are unlikely to see in binoculars the dark feature that gives it its name, but you may be able to see that it is slightly elliptical.

Finding the Virgo galaxies proper requires care. Rather than cast around wildly, it is best to start at one edge and move in by hopping from one star to another. The galaxies you are seeing will not jump out at you, and on a poor night they are hard to see even with a large telescope equipped with computer control. One popular route is to begin with Denebola in Leo's tail. Move 6½° east and find the fifth-magnitude star 6 Comae Berenices. Look also for the fifth-magnitude Rho Virginis to the southeast, which has fainter stars on either side of it. Place 6

▲ *The brightest galaxies in the Virgo Cluster are visible in reasonably dark skies. Look between 6 Comae Berenices and Rho Virginis for a triangle of stars.*

► *A sketch of the center of the Virgo Cluster as seen through 10 × 50 binoculars by Jeremy Perez. M86 and M84 are in the center, with M87 near the bottom. A faint object east of M86 and M84 is in the correct position for tenth-magnitude NGC 4468. However, the galaxy south of M86 is in reality two eleventh-magnitude stars, showing the difficulty of being certain what you can see at the limit of visibility.*

Comae Berenices in the corner of your field of view and look midway between that star and Rho. Near the middle of the field is a south-pointing equilateral triangle of eighth-magnitude stars. M87 lies near this triangle, and though you might not see it immediately, now you know where to look you may pick it up very close to an eighth-magnitude star. M84, M86, and M60 are also in the same area.

The brightest galaxy in the Virgo Cluster is M49, which lies just over 4° south of M87, probably within your same field of view. These are all elliptical galaxies, so they appear as little fuzzballs, with no structure. As is often the case, higher magnification is better than lower.

There is another galaxy in Virgo that is somewhat brighter and also considerably closer than the others. This is M104, also known as the Sombrero Galaxy from its shape in photographs. It is quite separate from the Realm of the Galaxies, and the easiest way to find it is start with the northwestern star of the Corvus quadrilateral and look for a line of seventh-magnitude stars ending in a triangle that appears to point to the galaxy. But beware—at the end of the line is a chain of three stars of eighth magnitude that are aligned almost east–west, like the galaxy itself. At low magnifications it is easy to mistake these for M104, which is some 22 arc minutes to the east and considerably smaller and fainter.

M104 appears elliptical and has an almost starlike core. With very large binoculars and clear skies you may be able to discern that one

▲ *The Sombrero Galaxy, M104, is well-signposted by the line of stars from Corvus but is quite low in the sky as seen from the northern hemisphere.*

edge of the ellipse is flattened, as a result of the dust lane crossing the galaxy which gives it its name.

Northern hemisphere only

Mizar and Alcor

The middle star of the tail of the Great Bear—or the handle of the saucepan, whichever you prefer—is Mizar, a favorite double star with many observers. Anyone with reasonable eyesight can see a fainter star nearby, named Alcor. These two stars were called "the horse and rider" by Arabic astronomers, and Arabic writers in the thirteenth and fourteenth centuries recorded that this pair was a test of the eyesight. One can only conclude that the eyesight test was designed not to be too stringent so as not to exclude potential job applicants, maybe, because it is an easy naked-eye object. The separation of the two stars is just under 12 minutes of arc, and Alcor is almost exactly magnitude 4. Mizar itself is a double, though as the separation of the two stars is 14.4 arc seconds, you will need a magnification of at least 10× to see the two stars separately and your binoculars must be very firmly mounted. There is a magnitude-7.6 star making a triangle with Mizar and Alcor.

Southern hemisphere only

The Jewel Box, NGC 4755

When Sir John Herschel referred to the cluster of stars around Kappa Crucis as "a casket of variously colored precious stones," he was doing a very good PR job on it. As a result it has become known as the Jewel Box Cluster, and everyone wants to see it. However, to the naked eye it appears as just a single fourth-magnitude star, which is why it was called Kappa Crucis, and it does not yield up its jewels readily. Most binoculars show this star, just to the east of Beta Crucis, as nothing more than a slightly enlarged triangular object. It is not until you use an instrument with a power of around 25× that you can really begin to see the individual stars properly, and see the red star that gives the cluster its popular name. The individual stars are about sixth magnitude.

The May sky

The bright star Arcturus dominates May skies. This is the fourth brightest star in the night sky, and the brightest in the sky's northern hemisphere, the others being Sirius, Canopus, and Alpha Centauri. It is noticeably orange in color as it is a red giant star. It is the main star in the constellation of Boötes, the Herdsman, which extends as a kite shape to its north. To its south lies Virgo, and Spica, and to its north is Ursa Major. The curved tail of Ursa Major makes a useful indicator: follow its curve and you reach Arcturus; continue it farther and you reach Spica. However, this is one of the most barren areas of the sky for good binocular objects, with nothing much of note in Boötes even when using a telescope. The name, incidentally, is pronounced "Bo-oh-teez."

In the far south, things become more interesting as we approach the Milky Way. The stars of Centaurus crowd the sky, with the brilliant Alpha and Beta Centauri pointing to the Southern Cross (Crux). To those of us who like star patterns, the remainder of Centaurus is a problem. There is no neat figure that we can identify with or build upon, but instead a collection of moderately bright stars.

▼ The Jewel Box NGC 4755 in Crux surrounds the star Kappa Crucis.

What's a galaxy?

Until the early twentieth century, it was not appreciated that many of the misty blobs in the sky are more than just tightly knit clusters of stars so distant that the individual stars could not be made out. In fact what we now refer to as galaxies are groups of not just hundreds but hundreds of millions of stars, very distant from our own Milky Way, which is itself our own Galaxy and is distinguished by being given capital letters. Actually, the words Milky Way can refer both to the luminous band in the sky and to the object as a whole, as seen from outside—though it is a fair bet to say that humans will never be able to travel beyond the Milky Way Galaxy. Not even the fictional starship *Enterprise* could do that.

Galaxies come in various shapes. The simplest is a spherical ball of stars, but many of this type are squashed spheres, or ellipses, and they are all known as elliptical galaxies even though they might happen to be spherical in shape. It is not easy to tell just by looking whether we are looking at a genuine sphere or one shaped like a football seen end on.

Our own is a spiral galaxy, with the beautiful spiral arms that we see so clearly in photographs of similar external galaxies, though being inside it we can't appreciate them. At the heart of a spiral galaxy is a nucleus which resembles an elliptical galaxy, with the rather flat disc that contains the spiral arms spreading out from the nucleus. Again, the appearance of a spiral galaxy changes with its orientation to our line of sight. We may see it face on, in which case the arms are spread out; or edge on, in which case it looks like a spindle, or somewhere in between. The nucleus is usually the brightest part, so often all you see of a spiral galaxy with binoculars is the nucleus, and so it is hard to tell a spiral from an elliptical. The spiral arms only show up well in long-exposure photographs.

There are also irregular galaxies, though the Magellanic Clouds are the only examples of these bright enough to be seen with binoculars. Irregular galaxies are generally smaller than spirals or ellipticals, and it seems that in the early

The Whirlpool Galaxy, M51

There are some objects that feature in virtually every book on astronomy, and M51 is one of them. A face-on spiral galaxy, it was the first object to be seen as a spiral, by Lord Rosse in 1845, using what was then the largest telescope in the world, a giant 1.8 m (72 in) reflector. The object itself had of course been seen many times previously, and featured in Messier's catalog, but Rosse was the first to see clear spiral structure. At the time, and indeed for many years afterward, the true nature of M51 was not known as the whole concept of external galaxies had yet to be proposed.

M51 is quite easy to locate, though from much of the southern hemisphere it is too low in the sky to be worth looking at. Start from Eta

Universe, all galaxies may have been irregular. Close encounters between neighboring galaxies caused shock waves that created the spiral arms, and mergers resulted in sudden bouts of starbirth that rapidly robbed the individual galaxies of their gas and dust, leaving the featureless ellipticals. In big clusters of galaxies, the largest galaxy is invariably a giant elliptical somewhere near the center.

▲ *Types of galaxy. M87 is elliptical, M83 is a face-on spiral and NGC 253 and M104 are edge-on spirals. NGC 5128 is a colliding pair of galaxies.*

Ursae Majoris, Alkaid, which is the end star of the Great Bear's tail, or the tip of the saucepan's handle. M51 forms a wide triangle with a fourth-magnitude star near Alkaid, 24 Ursae Majoris. The map on page 76 shows an outer circle of 5° diameter.

Don't expect anything large, but you should be able to spot a small elongated fuzzy patch. Luginbuhl and Skiff, in their *Observing Handbook* and *Catalogue of Deep-Sky Objects*, report two separate fuzzy patches using only 7 × 35 binoculars, the second patch being the companion galaxy, NGC 5195. Others, however, say that they have trouble finding M51 even with 10 × 50s on good nights. This may be because the galaxy really is quite small and easily overlooked if you are not accustomed to finding faint galaxies. However, do not expect to see the

◀ *A finder chart for M51 from Alkaid. The map has a field of view of 5°.*

spiral structure—this really requires larger apertures than most binoculars provide.

A sketch and photograph of M51 are shown on page 8.

Globular cluster M3

This object, like M51, lies within the constellation of Canes Venatici, the Hunting Dogs. It is certainly larger and brighter than M51, so if you have no luck with the galaxy, try M3, which is one of the largest globular clusters visible from the northern hemisphere. It is a bit isolated, and perhaps the easiest way to find it is to start from the neighboring constellation of Coma Berenices. The star Gamma Comae Berenices is the brightest in the Coma Star Cluster, at magnitude 4, and to its east is Beta Comae Berenices, of similar brightness. The two stars are 10° apart. Project the line a further 6° and you should spot M3 as a fuzzball, quite a lot brighter at the center (see map on page 68). With the lowest-magnification binoculars it might look somewhat starlike, particularly in light polluted skies.

Omega Centauri and NGC 5128

For observers in Europe, these objects are forever below the horizon, but from much of the southern US they are accessible if rather low down, so choose the clearest night and a good southern horizon. We treat them together because having found the brighter but more southerly Omega Centauri, NGC 5128 is easier to locate. The route to Omega varies depending on where you live.

From the northern hemisphere, take a diagonal through the quadrilateral of Corvus from top right to bottom left, and keep going twice as far again to find third-magnitude Iota Centauri. From there, continue the line to find a pair of third-magnitude stars, Mu and Nu Centauri. They point southward (with a little cheating) to third-magnitude Zeta Centauri, and Omega Centauri appears as an object of similar brightness to the west of this star. It appears starlike at first glance to the naked eye, but is clearly fuzzy when you look a bit more carefully.

From the southern hemisphere, you can use the above route or use the western and northern points of the Southern Cross (Crux) to point

▶ *Sketch of Omega Centauri as seen though 10 × 50 binoculars by Chris Picking of Wellington, New Zealand.*

to Omega. Project the line about four and one half times and you should arrive at Omega—it really does not take much finding.

From Omega, look just 4½° north—usually less than one binocular field—and you can spy a small fuzzball, which is the galaxy NGC 5128. This may not look very much in binoculars, but photographs show a spectacular galaxy in the throes of a collision with another. This is an intense source of radio waves, known as Centaurus A. You might think that the dramatic collision would wipe out any life forms existing on planets in either of the two galaxies, but the stars and planets would probably pass right by each other, such is the distance between individual objects.

Photographs actually give a misleading view in one respect, compared with your own eyes. The photographs make a galaxy look as if it is a solid wall of stars, but in fact this is because the light from each star spreads out on the pixels of the camera being used, as the light builds up. When you look at the galaxy, you can see that the surface brightness is actually very low. If it were a wall of stars, one object would overlap another and the result would be a dot of light with the same surface brightness as our Sun, a typical star. As it is, most lines of sight through the galaxy pass right through, or end in dark dust and gas. And even if you were close to the galaxy, it would not appear significantly brighter than our own Milky Way does to us.

M83

This is a bright galaxy which, being fairly far south, is difficult to view for many northern European observers. But from the USA, from much of southern Europe, and southern France southward it is worth looking for as it is easier to spot, under the right conditions, than M51. Having found Iota Centauri, as mentioned opposite, look for second-magnitude Theta Centauri to its east, then imagine an equilateral triangle to the north of the two stars and you should be in the right area. To help you further: look for a triangle of fifth-magnitude stars to the north of the line between the two stars. The two westernmost of these point north to M83, which lies slightly to the west of a sixth-magnitude

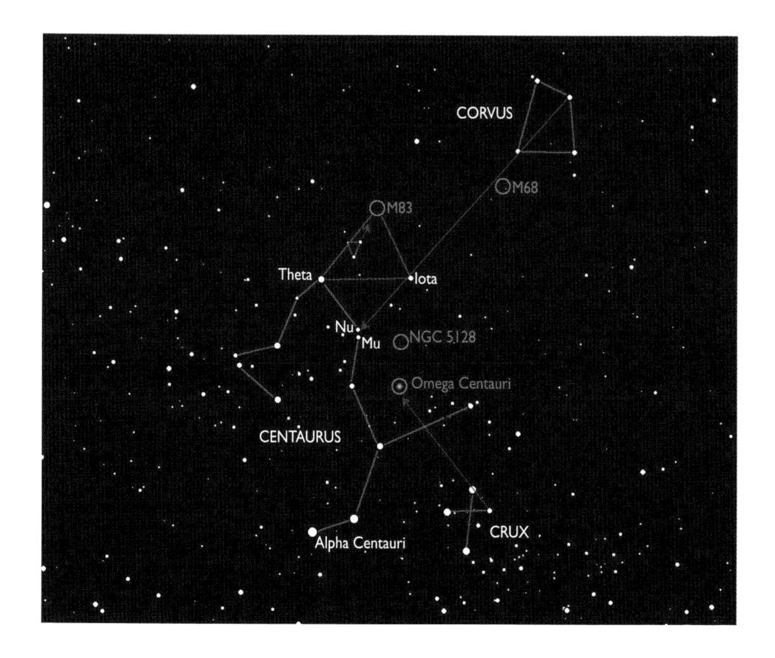

▲ *Your routes to the deep sky objects of Centaurus and Hydra from either the northern or the southern hemisphere. Crux is visible south of latitude 25°N.*

star. M83 is a face-on spiral galaxy, and while ordinary binoculars will not reveal the spiral arms, you may see that the center of the galaxy is more condensed than the outer regions.

While you are in the area, you might look for M68, which is a globular cluster found by projecting the line between the two easternmost stars of the quadrilateral of Corvus southward about half as far again.

Southern hemisphere only

Alpha Centauri

We can't let this part of the sky go without commenting on Alpha Centauri, the nearest bright star to the Sun at 4.39 light years. It is actually a double star, though the two components are quite close together. The two orbit each other every 80 years in quite elliptical orbits, so the separation between them changes from year to year. Currently they are closing to a minimum separation of about four arc seconds in 2015, which puts them beyond easy observation with binoculars. The brighter of the two stars is very similar to the Sun—about 10% larger and brighter—while the other is somewhat smaller and dimmer, and more yellowish.

The closest star of all, Proxima, orbits these two. It is 4.22 light years away, so you might be tempted to search for it. However, it is a red dwarf star and even though it is close, it is so faint as a star that it is only eleventh magnitude and is hard to find among many other background stars of similar magnitude. It lies some 2° away from Alpha.

The June sky

Like those of May, the June skies are fairly barren of stars until you get down to the far south and the Milky Way, and the constellation of Scorpius. For those observers in the central and northern Canada in particular, June skies are very light because the Sun is not far below the horizon, and it never gets fully dark at night. Astronomical twilight is defined as being when the Sun is below but within 18° of the horizon, and in much of Canada and northern Europe this never occurs during the whole of June and the first two weeks of July. So the June skies are best observed earlier in the year but later in the night.

The constellations of Ophiuchus, Libra, Serpens, and Hercules have few bright stars and little in the way of recognizable patterns. The most readily identifiable are Hercules, which has a central area known as the Keystone, consisting of third-magnitude stars, and the attractive small constellation of Corona Borealis, the Northern Crown, with its semicircle of stars. Europeans saw it as a crown, but the indigenous Australians saw it as a boomerang. The brightest star of the semicircle is a fairly respectable second magnitude. However, Scorpius is instantly obvious with red first-magnitude Antares flanked on either side by two third-magnitude stars, and with a curving pattern of stars that mark its tail, though the latter is below the horizon from northern Europe. Note—only astrologers call this part of the sky Scorpio!

The June skies are crawling with globular clusters. To the binocular observer these can be frustrating objects to track down, particularly with magnifications lower than 10×, because they can appear virtually starlike. However, with the higher powers they begin to show more character, and each one is subtly different. Globular clusters are all rather remote, and a telescope is really needed to reveal stars in any of them. However, they can be comparatively easy to see if you observe with care.

M13—the Great Hercules Cluster

Southern-hemisphere observers will scoff at the description of M13 as a great cluster, and it is not even the brightest or largest cluster visible from the northern hemisphere; but the fact remains that for many observers it appears the best because it rises high in the sky and is in an otherwise barren region. It is also easy to locate, being between the two western stars of the Keystone of Hercules. It is almost exactly two

About stars

We are fortunate that our own star, the Sun, is a stable and well-behaved star. Actually, good fortune has little to do with it. If it were in any way unstable, life would probably have not evolved on Earth as it has done, so we would not be here to talk about it. There is a good reason for the Sun's stability. It is a star of average mass, and is about midway through its lifetime.

When stars condense from starbirth nebulae, such as the Orion Nebula, their future lives are mapped out for them from the very moment of their birth. They shine by nuclear fusion reactions, which proceed more rapidly the hotter the star. Their mass determines pretty well how long they will live and how bright they will be. Stars much less massive than the Sun, such as the tiny Proxima Centauri, are red dwarfs and glimmer with only around one ten thousandth of the Sun's brightness. But they will last for many billions of years longer than the Sun.

A star of the Sun's mass can expect to use up its hydrogen fuel within about 10 billion years of its birth. The Sun is currently about halfway through its normal lifetime, as what is called a main sequence star. Such stars are converting hydrogen into helium, and their color is a good guide to their temperature and therefore the rate at which they will use up their fuel. The coolest stars are red, then come orange, yellow, white, and blue-white stars. These are just the same colors that you would get if you heated up a piece of metal to the same temperatures. The filament of an incandescent lamp is just such a heated metal, and an ordinary 100 watt light bulb is the same color and temperature as a red dwarf star, while hotter-running halogen lamps appear slightly bluer.

A star of much greater mass than the Sun, such as many of those in Orion, will stay on the main sequence only a matter of millions of years rather than billions, shining very brightly and with a bluish light.

When a star's supply of hydrogen fuel starts to dwindle, the energy production in its core is no longer enough to support the pressure of the material in the star, and it undergoes a drastic change. Its outer layers swell outward and its temperature goes down, but because its surface area is now much greater, it shines more brightly than previously. This is a red giant star, and such stars are noticeably orange or red. Arcturus is an example of such a star. Finally, maybe after a phase of variability, the star will collapse, puffing off its outer layers which expand away in a shell surrounding the star to become a planetary nebula. The core of the star itself becomes a very dim white dwarf, of which the companion star to Sirius is an example. This will be the ultimate fate of the Sun.

Much more massive stars are unable to rearrange themselves in this way, and undergo a catastrophic demise at the end of their brief period on the main sequence. To begin with they may swell to become red supergiant stars, Betelgeuse in Orion being an example. Eventually, they have no choice but to explode violently as a supernova. So much energy is produced in such an explosion that the star may briefly become almost as bright as all the other stars in a galaxy put together. The decimated remains of such a star collapse down into a neutron star or even a black hole. In a neutron star, material is incredibly tightly packed—the mass of a whole star bigger than the Sun can condense down to a body just 10–20 km (6–12 mi) across. Any spin that it previously had is enormously accelerated, so it rotates very rapidly indeed. In 2007, a neutron star was observed to have a rotation rate of a record 1122 times a second, for example, though the observations required confirmation.

The most massive bodies cannot support themselves under the gravitational crushing of their outer layers, and almost certainly become black holes, from which nothing can escape. However, despite what you may have seen in the movies, black holes don't go around sucking everything into them. They still have the same gravitational pull that they did when they were stars—actually less, as they will have lost mass in the supernova explosion. And although they can't emit light themselves, any material that falls into them will heat up as it plunges into them, so the surroundings of a black hole can be anything but black, and emit copious amounts of X-rays.

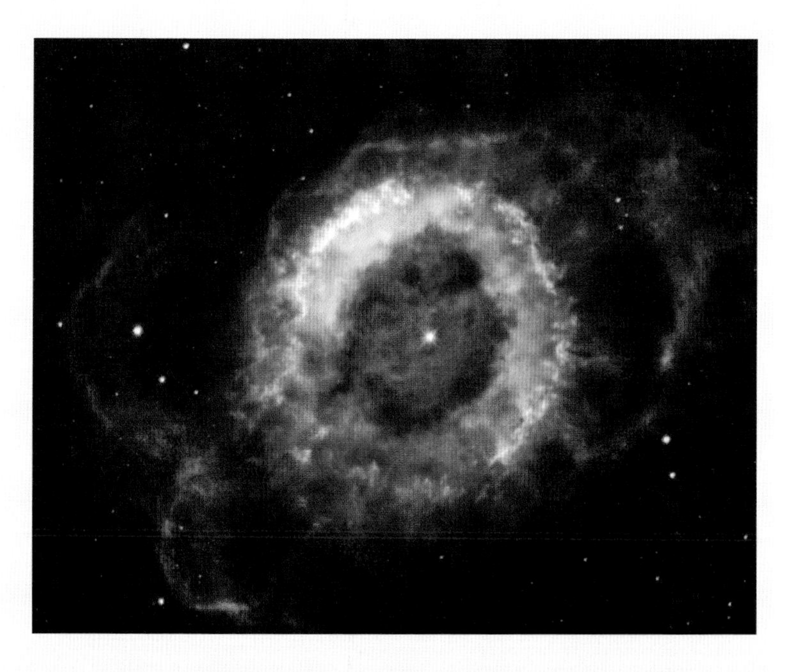

▲ *Planetary nebula NGC 6369, photographed by the Hubble Space Telescope. The vivid colors are signatures of the chemical elements released in shells by the star, though they are not easily seen by eye.*

▲ *This deep photograph of M13, taken from Essex, UK, has north top left. The two stars flanking the cluster are just visible to the naked eye.*

thirds of the way from Zeta to Eta, and is flanked by two seventh-magnitude stars. These help to augment its brightness as seen with the naked eye, so although the cluster is actually on the verge of naked-eye visibility, many people can spot that there is something there because of the combination of three objects rather than just the one.

Binoculars with a magnification of 7× show it as slightly non-starlike, though the darker your skies the better, while by the time you get to magnifications of 12× or 15× you can begin to see that it is a condensed fuzzball, brighter at the center. It actually consists of some half million or even one million stars within a space of about 85 light years.

M92—poor relative

Having found M13, the next port of call is nearby M92, which is slightly less bright and more condensed, so you may have more trouble with it. It forms a triangle with the northern edge of the Keystone, though there is a fifth-magnitude star that forms a flatter triangle which you should not confuse with M92 itself. It is reputedly the most ancient globular cluster in the Galaxy, with stars that are as old as the Universe itself. This globular makes an interesting comparison with M13, but is overshadowed by its near relative.

R CrB: now you see it...

As a change from globular clusters, take a look within the semicircle of Corona Borealis. You will probably see a star of sixth magnitude within it, as shown on the map. This is R Coronae Borealis, known to its friends, by the abbreviation of the constellation, as "R CrB." For months or even years on end this star shines steadily at about magnitude 6, but suddenly, without warning, it will disappear from view. Investigation shows that it has dropped in brightness by several magnitudes, maybe down as low as magnitude 14. Over a matter of months it climbs back to its normal state.

When it fades, the star has puffed off a shell of material that hides it from view, but this dissipates over a period of time. Exactly why and how this happens is uncertain, and it is often amateur observations that alert professionals to its sudden changes.

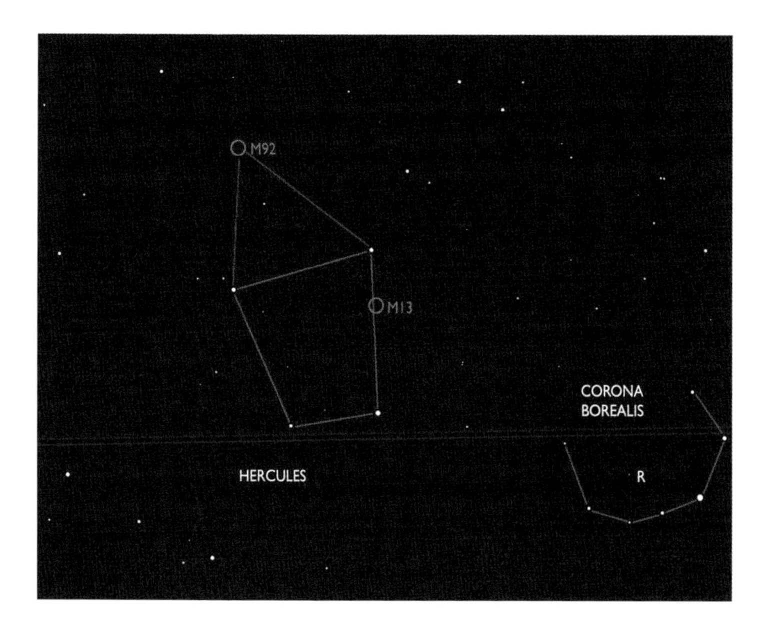

▲ *Finder chart for the globular clusters M13 and M92 and the disappearing star R Corona Borealis. The Keystone of Hercules is marked.*

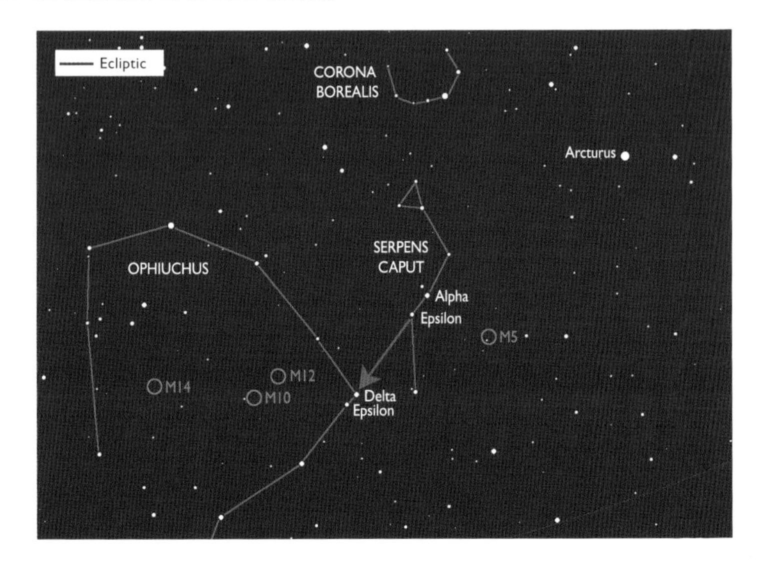

▲ *Start at Corona Borealis to star-hop to the four globular clusters M5, M10, M12, and M14. Ideally, use binoculars of 10× or more.*

Overlooked M5

There is some controversy over M5. It has been called the brightest globular cluster in the northern hemisphere, yet most people would give that title to M13. It is somewhat lower in the sky than M13—though from even northern Canada or Europe it is at a respectable altitude—but it is definitely harder to locate. Catalogs show it as being magnitude 5.7, compared with 5.8 for M13. So what's going on?

The answer is that the collective brightness of all the stars in M5 is greater than that of M13, but the cluster is somewhat larger in the sky. As a result, it has a slightly lower surface brightness. So if you have a good sky, you could actually find M5 a more rewarding object than M13. The answer is to take a look for yourself. One way to find it is to begin at Corona Borealis and locate the stars of Serpens Caput—the head of the serpent. This poor creature is in two halves in the sky—its tail, Serpens Cauda, is down near Sagittarius. A triangle of three fourth-magnitude stars to the south of Corona Borealis marks the actual head, while Alpha Serpentis is a second-magnitude star an equal distance south of this, with two other stars close by making a triangle. Some way to the south of this is an east–west line of stars which continue into Virgo, one of which is the fifth-magnitude 5 Serpentis. Your target lies close to this star, to its northwest. As always with globular clusters, look for a fuzzy star rather than a group of stars.

M10 and M12—globular twins

Two globulars that are easy to compare, M10 and M12, lie within the same binocular field of view in Ophiuchus. Find them by going to Alpha Serpentis and following the line pointing southeast indicated by Epsilon Serpentis until you get to an obvious pair of third-magnitude stars, Delta and Epsilon Ophiuchi. Now look two binocular fields (about 9°) to the east, and you should spot M10 and M12, with M10 being the brighter and more southeasterly. Not only is M10 the richer of the two, it is also closer, at about 14,300 light years compared with 16,000.

Having found M10, move exactly 10° to the east and you may pick up the considerably fainter M14, which is about twice the distance of M10 at 30,000 light years.

Antares, M4, and M80

For another contrasting pair of globular clusters, look south to Antares, the brightest star in Scorpius, which features in some of the most beautiful astronomical photographs ever taken. Antares itself is easy to spot, with its two guardian stars Sigma and Tau Scorpii. Long-

▲ *A photograph by Michael Stecker of the region from Antares (bottom) to Rho Ophiuchi (top). Sadly, views like this are restricted to long-exposure photographs.*

exposure photographs of the region show an astonishing variety of colors, with Antares itself casting an orange glow over the surrounding dust, which reflects the color of this red giant star. To its north lies Rho Ophiuchi, which lights up a surrounding blue reflection nebula. Sigma Scorpii ionizes a nearby cloud of hydrogen gas which glows deep red. Dust throughout the area creates a brownish tinge.

To one side of the line between Antares and Sigma is another fuzzball, this time the globular cluster M4. But this is a good example of how a photograph can be so different from the way the eye sees stars. Each star is a great blob, apparently far bigger than M4. Look through binoculars, however, and you will see that Antares and Sigma appear as points of light, while M4 alone is fuzzy. The imaging system, whether film or digital, causes the light from a star to spread out as soon as the immediate spot is saturated with light, whereas your eye can see a far greater brightness range than any photograph can reproduce. Sadly, none of the brilliant colors are visible using binoculars, though Rho can be seen in binoculars as a pretty triple star.

There is another globular cluster in the area—M80—which forms a triangle with Sigma Scorpii and Rho Ophiuchi, but it is faint compared with M4, which is one of the brightest globulars visible from the northern hemisphere and also probably the nearest globular to the Sun at about 7200 light years. Even with binoculars, it starts to look like a genuine globular cluster rather than a fuzzy star.

NGC 6231—table or comet?

The great curve of stars to the southeast of Antares is what creates the appearance of a celestial scorpion, which is how this region has been regarded since Mesopotamian astronomers first wrote down the names of the constellations. The point where the sting begins to curve eastward is the location of a very condensed cluster known enigmatically as the Table of Scorpius. In most binoculars this object, NGC 6231, is not well resolved and appears as an elongated haze, but taken together with a fan shape of stars to its north, known as Trumpler 24, the whole area has the appearance of being a comet with a tail. Had Messier been able to see this far south from France he would undoubtedly have included this in his catalog, but it is invisible from central Canada and frustratingly low from much of the northern US and southern Europe.

Northern hemisphere only

Ursa Minor and Polaris

The Pole Star is not the brightest star in the sky, but some might argue that it is the most important, because it lies so close to the north celestial pole. Navigators have used it as a guide to the direction of north for hundreds of years, and a measurement of its distance above the horizon gives you an instant reading of your latitude. For example, at the equator, latitude 0°, it is virtually on the horizon, whereas from New York, at latitude 40¾°, it is at that angle above the horizon. It lies at the end of the tail of Ursa Minor, the Little Bear, which has a box for a body and a curved line of stars for a tail. Both celestial bears have long tails, unlike their earthly cousins.

The stars of Ursa Minor are useful for gauging the faintest star you can see, known as the limiting magnitude, as they are all of about fourth and fifth magnitude. Use the map on page 88, which shows magnitudes of selected stars, to judge for yourself. As the area always remains at the same altitude above your horizon, you can be fairly sure that the results you get are a true measure of the limiting magnitude of your site.

With binoculars, Polaris itself appears just like any other star. But if the skies are good enough you can see a small circlet of stars to one side of it. This is known as the Engagement Ring, and has Polaris itself as a diamond in the ring.

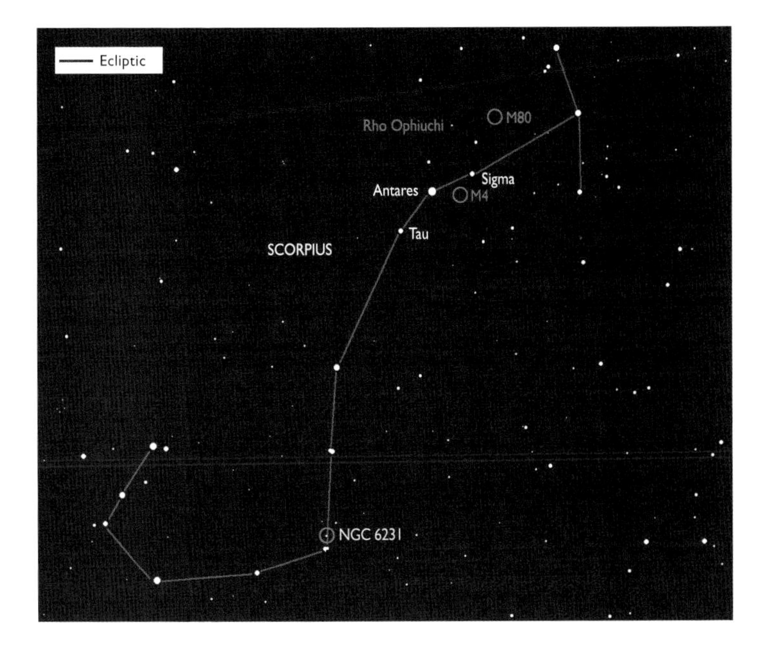

▲ *Scorpius and its binocular objects. In addition to the stars on the map, a bright planet may be visible along the red line, which marks the ecliptic.*

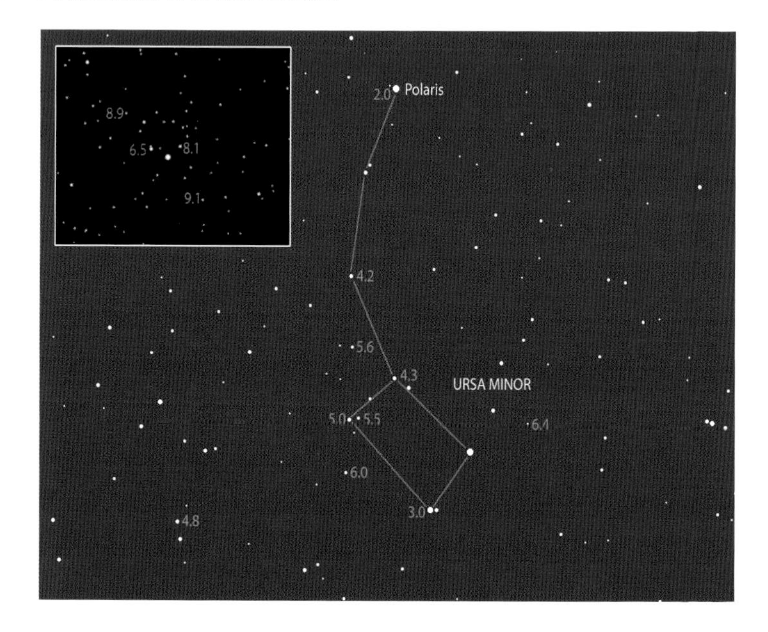

▲ *Find your naked-eye and binocular limiting magnitude by looking at Ursa Minor. The inset shows an area of 3° around Polaris.*

Southern hemisphere only

NGC 6397—close rival

The fourth brightest globular cluster in the sky is in the constellation of Ara, the Altar, which despite being far south was one of Ptolemy's original 48. Ara looks rather more like a high-backed chair than an altar, at least as seen from the northern hemisphere, consisting as it does of two slightly curving lines of stars. Your target makes a flat triangle with Alpha and Beta Arae—look for the usual fuzzy star. But this one looks distinctly more like a fuzzball than many other globulars, because it is not very condensed toward the center. NGC 6397 is second only to M4 in being the closest globular cluster to the Sun, and since these measurements are notoriously inaccurate, it may even be the closest.

NGC 6067 and NGC 6087

The constellation of Norma is not well known, nor does it consist of particularly bright stars, but it contains a couple of good open clusters, NGC 6067 and NGC 6087. To find them it is best to begin with Alpha and Beta Centauri. In the opposite direction from the Southern Cross they point to a pair of fourth-magnitude stars, then a roughly equal dis-

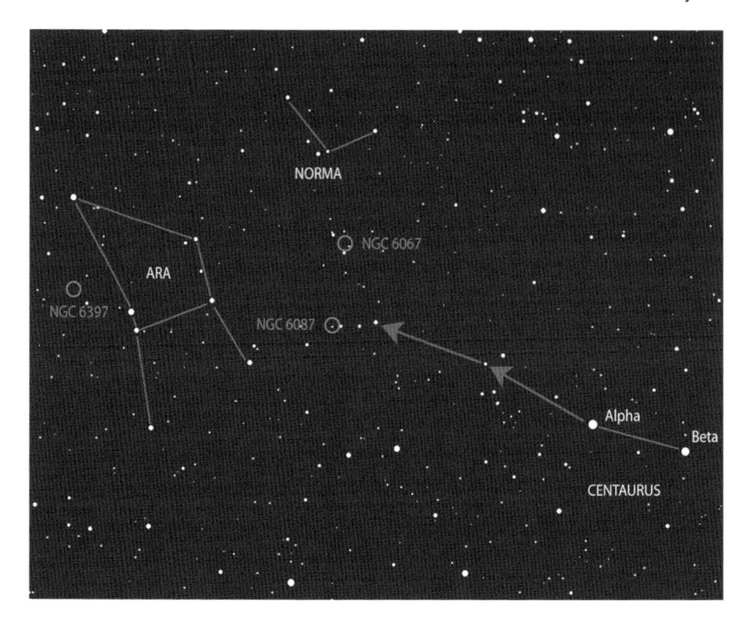

▲ *Use the Pointer stars, Alpha and Beta Centauri, in the reverse direction from usual to locate binocular clusters in Norma and Ara.*

tance farther on to a little line of stars of decreasing magnitude, at the end of which is NGC 6087, a haze of stars mostly fainter than eighth magnitude. From here, move 3½° northward to spot the haze of NGC 6067. Small binoculars are hard pressed to show individual stars, as the majority are between magnitudes 9 and 11. There are outlying stars that enhance the appearance of this cluster though they may not be actual members.

The July sky

The equatorial part of the July sky is a bit nondescript. The constellations of Hercules and Ophiuchus, the Serpent-Bearer, occupy most of it. Draco, the Dragon, lies to the north, with a parallelogram of stars marking its head. Vega, one of the brightest stars in the sky, we will come to next month.

Farther south, however, Sagittarius is a prominent constellation, and features the brightest part of the whole Milky Way. From southerly latitudes in particular this region is high in the sky and is bright enough to be obvious even in suburban areas, though from the northern hemisphere it is lower and less spectacular. To many people the main stars of Sagittarius look like a teapot pouring hot water onto the tail of the

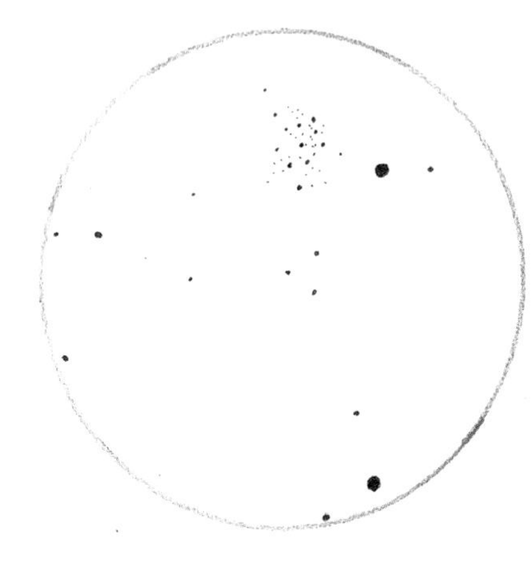

▲ *Eddie Horsley's sketch of IC 4665 as seen through 10 × 50 binoculars.*

Scorpion. Despite lying in the direction of the center of the Galaxy there are no really bright stars here—all the stars of the Teapot are of second or third magnitude. Sagittarius contains more deep sky objects than any other constellation.

IC 4665 and NGC 6633—forgotten clusters

It is often the case that the most popular areas of the sky are those with the most easily recognized constellations. Ophiuchus is a case in point. Its stars are widely scattered and do not have an obvious pattern. As a result, some of its deep sky objects are less frequently observed than if they were in other, better-known areas. To find two of these "forgotten" clusters, first locate Alpha Ophiuchi, also known as Rasalhague, a second-magnitude star to the south of Hercules and well to the west of Altair. The red star Rasalgethi, Alpha Herculis, lies about 5° to its west and a little north.

Some 8° south of Rasalhague and a little east lies the third-magnitude Beta Ophiuchi, which also goes by the name of Cebalrai. Just to the north and east of this star, and in the same binocular field, you will find the scattered stars of IC 4665. At around seventh magnitude they are easy to find even in light polluted skies, and as they cover a degree or so of sky they are best viewed with binoculars rather than a telescope. Now move 10° farther east and look for the smaller and fainter NGC 6633, just to the north of a sixth-magnitude star. Its main stars are magnitude 7 and 8, and it is a much better binocular object than quite a few better-known objects in the more popular tourist areas.

M11—the Wild Duck Cluster

A good name always helps a cluster's popularity, and the Wild Duck Cluster, M11, is definitely one of these. To find it, start with Altair and move southwestward to a semicircle of stars which is partly in Aquila and partly in Scutum. It looks like a squatter version of Corona Borealis. Just south of this lies M11, which binoculars show only as a misty patch because this is a quite remote cluster. Most of the stars are around tenth magnitude, which is beyond the reach of most binoculars, but together they are equivalent to about sixth magnitude. The name comes from a rough V shape of stars within the cluster, supposedly resembling wild ducks in flight, but you really need a telescope to see this.

M6 and M7

If you are viewing from sufficiently far south that this part of the sky is well above the horizon, and have dark skies, there is an area to the north of the tail of the Scorpion that immediately looks interesting. You can see hazy patches here with no difficulty, making this a prime target for binoculars. You will spot two large and bright clusters, M7 being much the more prominent with its scattering of fifth- and sixth-magnitude stars. M6, however, has a larger number of fainter stars, mostly between eighth and tenth magnitude, though its brightest star, at magnitude 6, is an orange giant star. The difference in appearance is largely due to the different distances of the pair—at some 800 light years, M7 is about half the distance of M6.

You should be able to see individual stars in both clusters with virtually any binoculars. These are two of the showpieces of the sky, though they are very difficult if not impossible to see from most of Canada or northern Europe—from even the northern US you will need a perfectly clear and dark horizon.

▲ *July's star clusters require some careful star-hopping, starting from either Rasalhague in Ophiuchus or Altair in Aquila.*

The Sagittarius deep sky

There are so many deep sky objects in Sagittarius that you can hardly miss them, particularly if you are observing from far enough south for the constellation to be fairly high in the sky. One in particular grabs the attention with the naked eye if the skies are fairly good. There are few nebulae that are easily visible in binoculars, let alone the naked eye, and the Lagoon Nebula, M8, is one of them. Take a line from the top of the handle of the Teapot through the star at the top of its lid, Lambda Sagittarii, and keep going as far again, and you come to it. It appears as a small line of stars surrounded by an oval misty patch, but in this case the mist is real—no matter how much it is magnified with a telescope, it always remains misty. Its pale glow is bright enough to shine through a moderate amount of light pollution. Though it is by no means as bright as the Eta Carinae Nebula or the Orion Nebula, it is probably third on the list. Like the other bright nebulae it is a starbirth region, and some of the newly born stars are visible apparently intermingled with the gas.

Before moving on, look just to the northeast of Lambda and you should easily spot M22, for many the brightest globular cluster visible from the northern hemisphere. To northern European and Canadian

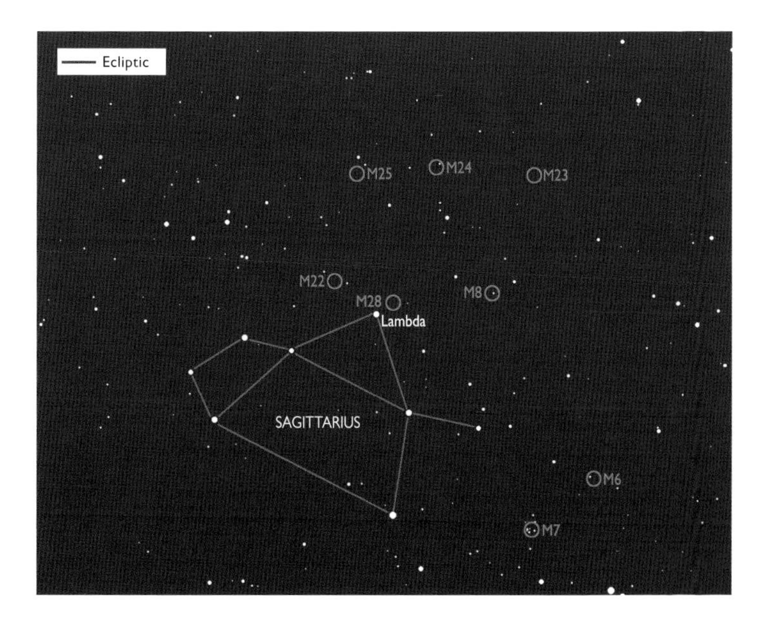

▲ *The Teapot shape of Sagittarius is your guide to the binocular deep sky objects in this area, including M6 and M7 in Scorpius.*

▲ *The splendors of the Lagoon Nebula as seen by the 158 inch (4 m) Mayall Telescope on Kitt Peak, Arizona. The nebula's name derives from the dark lane, which was seen as reminiscent of a lagoon surrounding a coral island.*

observers it is quite low in the sky, which may account for the fact that M13 is often feted as being the brightest. It would be better known if it were in a less crowded area—it is in fact the third brightest in the sky. Even small binoculars show this fuzzball clearly, as it is getting on for the apparent size of the Moon and you get the real impression of a globular cluster. You may also be able to spot the smaller and fainter M28, closer to Lambda and to its northwest, and within the same field of view of most binoculars.

Turn some 6° northward from Lambda and M22, and you come to a line of objects running east–west. The cluster M25 is an easy binocular object with its dozen or so stars of seventh to ninth magnitude and a haze of fainter ones. Within the same field of view to its west is an exquisite area variously known as M24 and the Little Sagittarius Star Cloud. This oval patch, some 2° across, is so rich in stars that you constantly feel that there are yet more stars to be seen. Like many lesser clusters, there are many stars in this direction that are too faint to be seen directly yet which add to the overall glow. Move a few degrees west again and you come to the third in the line, M23, which has fainter stars than M25, yet is still a superb sight on a good night.

Returning to M24, within a field of view to its north you find the Omega or Swan Nebula, M17. This is fourth on the list of the brightest nebulae in the sky. Smaller than the Lagoon Nebula, it is more fugitive in a light polluted sky or when at low altitude, but is nevertheless an easy object in the right conditions. You need a higher magnification than most binoculars provide to be able to see clearly the swan-like shape that gives it its name.

August skies

These are skies of summer for the northern hemisphere and of winter for the southern. Three bright stars dominate the sky as seen from most parts of the world except for the more southerly regions of Australia and New Zealand—Vega, Altair, and Deneb. Deneb is the most northerly of the three and is always below the horizon from south of about latitude 42°S. These stars are often referred to as the Summer Triangle, and like other celestial signposts they are very useful for pointing the way to other patterns. The Milky Way runs right through this part of the sky.

Even if you live in a light polluted part of the northern hemisphere, you can make out the line of the Milky Way by the greater number of fainter stars along its track. It runs from Deneb in the north past Altair to the constellation of Sagittarius in the south. In darker skies you can easily see a darker band along its length. This is known as the Great Rift and is caused by an immense molecular cloud several thousand light years away, which hides the stars beyond.

You hardly need a guide to find star clusters in this part of the sky. Just sweep along the line of the Milky Way with binoculars and small clusters, patterns, and chains of stars will spring into view.

Vega is one of the brightest stars in the sky, at more or less exactly magnitude 0. From much of the northern hemisphere it is virtually overhead on July evenings. The small but neat constellation of Lyra, the Lyre, surrounds it. To its east is Deneb, at one end of what is known as the Northern Cross but which is officially the constellation of Cygnus, the Swan. It works well in either context, as the cross shape also resembles a long-necked swan flying down the Milky Way with outstretched wings. Deneb marks its tail. The third star of the Summer Triangle,

Altair, is easily identified by the two third-magnitude stars flanking it, a group which resembles Antares and its companions. There are other small but attractive constellations in the area, such as Sagitta and Delphinus, which are easily recognized despite their small size.

The faint zodiacal constellation of Capricornus may be graced by a planet at times, but otherwise there is a distinct lack of bright objects until you get to second-magnitude Alpha Pavonis, in the constellation of Pavo, the Peacock. This is curious for being the only bright star with an English proper name—it, too, is called Peacock, as it needed a name in the 1930s when the British Royal Air Force published an air almanac.

Albireo

One of the sky's most celebrated double stars, this third-magnitude star marks the head of the Swan. Its two components are separated by 35 arc seconds, which makes it fairly easy to separate the two using even low-powered binoculars. But you might wonder why there is a fuss about it, unless you can view it at a higher magnification. At powers of 20× or more, the colors of the two stars become more obvious. The brighter of the two is a yellow star while the fainter is blue. If you ever doubted that stars have colors, the close proximity of these two should settle the matter.

The sky's coathanger

For many people, this little group of stars is a real favorite. Known officially as Collinder 399 or Brocchi's Cluster, it is known universally as the Coathanger. Seek it out with binoculars 8° south of Albireo and you will see why—it is a perfect little coathanger in shape, albeit upside down as seen in binoculars from the northern hemisphere. It is strictly speaking an asterism—that is, a pattern of stars—rather than a true cluster, as the individual stars are at widely differing distances. The main stars are all of around sixth magnitude, so for most people this is not a naked-eye object.

M27—the Dumbbell Nebula

Most planetary nebulae are too small to be visible with binoculars. M27 is one of the exceptions. The brightest guide stars in the area are those of the pretty constellation of Sagitta, the Arrow, which really does look like an arrow, about 5° east of the Coathanger. The Dumbbell forms a right-angled triangle with Sagitta's stars. It's visible as a small circular hazy glow, which only in telescopes appears to have the double shape that gives the object its popular name. At magnitude 7.4 it should be a fairly easy binocular object in country skies. While you are in the area,

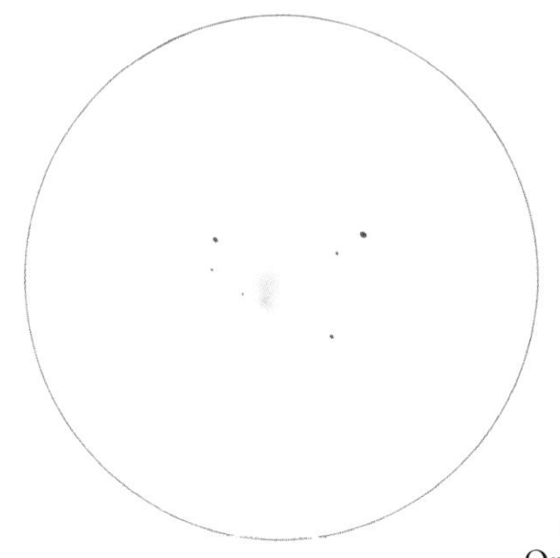

◀ *M27 as seen through 16 × 70 binoculars, sketched by Michael Hezzlewood from Burnley, UK.*

take a look for the cluster M71 between the two brightest stars of Sagitta. This is a cross between a globular cluster and an open cluster, with authorities tending to opt toward it being a globular these days.

North America Nebula: NGC 7000

One of the showpieces of the sky in photographs, and regarded by many people as a test of their observational prowess with binoculars or the naked eye. Yet others struggle to see anything here at all. The amazing resemblance to the continent is not nearly as obvious to the eye, no matter how good your instrument, as it is to film or a digital sensor. To digress, a digital camera gives a more accurate view of this area than does film, because the infrared filter that most digital cameras use to correct for their red sensitivity also renders them less sensitive to the red color of glowing hydrogen. So whereas film will easily show the distinct red nebula, a digital camera shot simply shows the star clouds which are more easily visible with the eye.

To see for yourself, look just to the east of Deneb. The center of the nebula is about 3° east of the star, so it should be within the field of view of most binoculars. Those who have seen it say that the "Gulf of Mexico" area is distinctly visible, whereas others see nothing more than a brighter area of the Milky Way. For once, low-power binoculars may have the advantage, as you need a wide field of view: the whole nebula covers over 2° of sky.

Using an astronomical light pollution filter could enhance the nebula's visibility. These are intended to be used in the eyepieces of telescopes, but if you happen to have one handy you can hold it behind the eyepiece of binoculars to try its effect, though they are not inexpensive items. Celestron make binoculars with built-in broadband light pollution filters for just this purpose. They do darken the field of view somewhat as they cut out the most offensive light pollution, but sadly the more light pollution you cut out, the less light there is left to view with. Broadband filters cut out just some of the light, and can be used on a wide variety of objects, but another type, narrowband filters, cut out all light except for a few specific colors emitted by nebulae in particular and have very specific applications.

▲ *Finder chart for August objects.*
▼ *The North America Nebula shows up clearly to the east of Deneb in a photograph made on red-sensitive film but is much more difficult to glimpse by eye, even with powerful binoculars.*

▲ *Barnard's "E," photographed from Kelling Heath, Norfolk, UK. The dark nebula is faintly visible to the right of the star Tarazed, to the northwest of Altair.*

Barnard's dark "E"

There are few dark nebulae in the sky that are easy to see from more northerly latitudes, apart from the Great Rift in the Milky Way which is visible with the naked eye, but one in Aquila is worth looking for with binoculars in good country skies as long as the Milky Way is visible. It lies just to the west of Tarazed, the most northerly of Altair's flanking stars. Gaze at this area with binoculars and you may be able to make out an absence of stars with the definite, if slightly ragged, shape of a letter E. This dark nebula was given the catalog numbers B143 and B144 by the great American observer E E Barnard (1857–1923), so the shape of the object is very fitting. Though you do need good skies to see this, you don't need to wait for that trip to the desert: it was obvious to author Robin Scagell from the annual Kelling Heath Sky Camp in Norfolk, England, in 2006.

Two doubles

The stars Alpha and Beta Capricornii make a distinctive pair, because each is a double star. Alpha, the more northerly, is easy to separate with the naked eye, but you may struggle with Beta, whose separation is 3.5 arc minutes. Binoculars, however, make it an easy target. You can see the two stars above Uranus in the photo on page 136 and the line of stars that includes Altair points south toward them.

The September sky

The Milky Way is now a feature of northern skies, running northward from Cygnus into Cepheus. But farther south the skies get very barren. The great Square of Pegasus dominates what is otherwise a very empty part of the sky. It is a rough square of second-magnitude stars some 15° apart that is only obvious because there are no other bright stars apart from first-magnitude Fomalhaut some 45° to its south. Moving south from there are two second-magnitude stars in Grus, the Crane (the bird, not the machine).

Aquarius occupies the equatorial regions of the sky, and like Capricornus it lacks bright stars so from many light polluted areas it is virtually invisible.

▲ *Cygnus is your starting point for locating interesting stars and binocular objects in September. The Milky Way runs through this area.*

Variables and doubles

Single, well-behaved stars such as our Sun are comparatively unusual. What appear to be individual stars are very often actually double or binary stars, with two individuals orbiting around each other. If you imagine twirling a dumbbell around, you will appreciate that the two objects actually orbit around a common center of gravity, and if one object is twice as massive as the other the point of rotation will be displaced toward that object. The separations between the stars can range from very close, with distances comparable with those between the Sun and its planets, and with orbital periods of just days, to very distant, in which case the orbital periods can be thousands of years. Some stars are even multiple, usually with close pairs of stars orbiting around other close pairs.

M39—a binocular cluster

To the northeast of Deneb is a pair of fifth-magnitude stars. Keep going past these an equal distance in a straight line from Deneb and you come to a fairly sparse cluster, M39, at the end of a ragged line of stars. Although there are only about a dozen stars in this cluster, they are all around seventh or eighth magnitude, so this is a good object to search for in light polluted areas and it is much better in binoculars than with a telescope because it is about the apparent size of the Moon.

Mu and Delta Cephei—jewels in Cepheus

If you want to see a really red star, take a look among the scattered stars of Cepheus, a constellation which is not blessed with a readily recognizable pattern. Alpha Cephei, Alderamin, is a second-magnitude star lying roughly midway between Deneb and the Pole Star, with third-magnitude Eta Cephei to its west. Just under 5° to the southeast of Alderamin lies fourth-magnitude Mu Cephei, a red supergiant star that is regarded as the reddest star visible with the naked eye in the northern hemisphere. Binoculars bring out the color particularly well. It varies in brightness somewhat over an irregular period.

Having found Mu, look 5° to its east to find a distinctive little triangle of stars. The easternmost of these, Delta, is very famous in astronomical circles as being one of the first and most important variable stars to be discovered. Its variations in brightness are not extreme—over a matter of 5 days and 9 hours it varies from magnitude 3.5 down to 4.4. But the crucial thing about it is that stars of this type, with the same distinctive pattern of variability, take longer to vary the brighter they are in real terms.

They are named Cepheid variables after this very star, and once a star has been identified as being a Cepheid and its period of variation

▼ *The "light curve" of Delta Cephei, formed by combining observations made over a year by members of the UK's Society for Popular Astronomy. The slow fall followed by a more rapid rise is characteristic of Cepheid variables.*

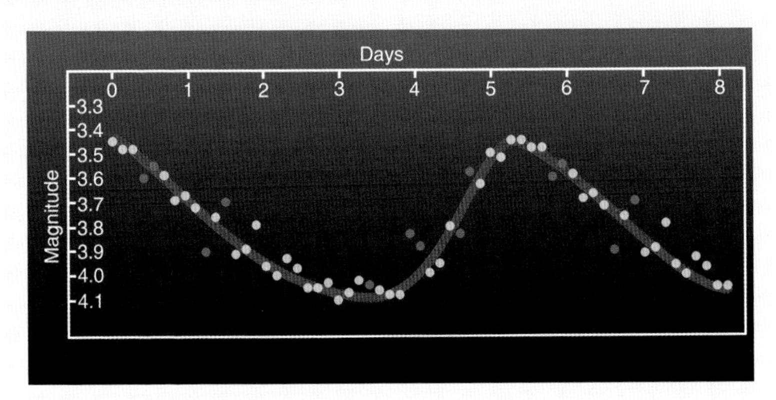

Many of the stars which we see as doubles in the sky are true pairs of stars, though usually if they are wide enough apart to be easily seen as individuals in binoculars their orbital period is likely to be so long that changes are not great even over a lifetime of observing. There are, however, some examples of apparent double stars where the two components are nothing to do with each other, and it is just chance that they are in line of sight. Alpha Capricornii is an example of this sort of double star, known as an optical double, with its individual components being 109 and 687 light years away from the Sun. The two stars of Beta Capricornii, however, are estimated as 314 and 344 light years from the Sun, so they may be a genuine pair. Though it sounds as if they are separated by 30 light years (compared with just 4.3 light years between the Sun and Alpha Centauri, which are not regarded as a linked pair), the uncertainties in each measurement are about 30 light years.

There are many different types of star whose light varies, known as variable stars. Some, such as Delta Cephei, are pulsating as they pass through a phase of instability. Yet others may flicker irregularly, or may have large starspots on their surface that come and go as the star rotates, or may be members of pairs with a star that is losing mass as it ages. The gas is dumped onto the surface of the star by its companion and at a certain point there is a flare-up, as you would get when throwing fat onto a fire.

Yet another type of apparent variable is caused when two stars are orbiting each other with their orbits almost edge-on to our line of sight. When the dimmer star hides the brighter one, we see what amounts to an eclipse, so these are known as eclipsing binaries. Algol in Perseus is the classic example.

measured, its true brightness is known quite accurately. Comparing this with its apparent brightness gives its distance—a notoriously difficult quantity to measure accurately, particularly for more distant stars. In fact, one of the reasons why the Hubble Space Telescope was built was to seek out Cepheids in galaxies in the Virgo Cluster, so as to improve our knowledge of its distance. This is an important stepping stone in our understanding of the distance scale of the Universe as a whole.

A lesser known feature of Delta Cephei is that it is a double star, whose components are separated by 41 arc seconds, so they are easily seen separately using most binoculars. The brighter star is yellow and the fainter blue.

Compare and contrast: M15 and M2

Though this object is in Pegasus, it is easy to find by starting from the easily spotted Delphinus. About 17° to the east of this constellation is a second-magnitude star, Epsilon Pegasi or Enif. About 4° to the northwest of this is M15, a fairly bright globular cluster, which itself lies close to a sixth-magnitude star.

Having found this, move 13° south and you should find another globular, M2. As an additional guide, this is just less than 5° due north

▲ *How to find two September globular clusters, M15 and M2, starting from the distinctive little constellation of Delphinus to the east of Altair.*

of the third-magnitude star Beta Aquarii. The total brightness of M2 is given as 6.6, while that of M15 is 6.3, but you may find that M2 is easier to see in a poor sky because it is more concentrated toward the center. No two globular clusters are the same, and if you are making sketches this is a good opportunity to compare the two.

The October sky

The W shape of Cassiopeia is high in the sky as seen from the northern hemisphere, while in mid-sky, visible from all parts of the world, the Square of Pegasus remains a prominent feature. You can use it to find two pretty asterisms. One, directly to its south, is known as the Circlet. It actually marks the head of one of the two fishes of Pisces, and consists of fourth- and fifth-magnitude stars so it is rather obscure from light polluted areas. The other, which you find by following the Square's diagonal from northeast to southwest an equal distance again, is the Water Jar of Aquarius. This arrow shape of stars of magnitude 3 and 4 is a distinctive feature of this part of the sky, even though these are not the brightest stars of Aquarius. In mythology, it marks the jar from which the mythical Aquarius constantly pours water, and all the constellations in this part of the sky have a watery connection.

▲ *The Circlet (left) and Water Jar (right), visible to the south of Pegasus. The extreme eastern star of the Circlet is a very red star, TX Piscium, whose color is evident in binoculars. The star varies between magnitudes 4 and 6.*

The Square of Pegasus shares its northeastern star with Andromeda, whose brightest stars are a widely spaced line of three, running at an angle to the northern edge of the Square. Everyone has heard of the Andromeda Galaxy, but in fact October skies contain several other nearby galaxies.

Andromeda galaxies

The Andromeda Galaxy, also known as M31, is virtually a twin of the Milky Way, located some 2.5 million light years away. The two are the leading members of our Local Group of galaxies, and each is inclined at an angle to the other. Compared with most of the other galaxies visible with binoculars, Andromeda is really easy, and it can just be seen from even the center of a large city on a good night using binoculars, and with the naked eye on average nights elsewhere, though you may need averted vision to spot it in less than really dark conditions.

To find it, begin with the Square of Pegasus. The rule is "count two along, and two north," the second star along being Beta Andromedae, at magnitude 2, and the others referring to third- and fourth-magnitude stars. Close to the star Nu Andromedae lies your quarry—a quite large and noticeably oval hazy blur. Its size depends on how dark your skies are. M31 is a full 3° in length, though in poor skies you may see only the central ¼° or less. While gazing at this pale blur, you can speculate that its inhabitants, whatever they may look like, can see our own Galaxy looking remarkably similar in their own

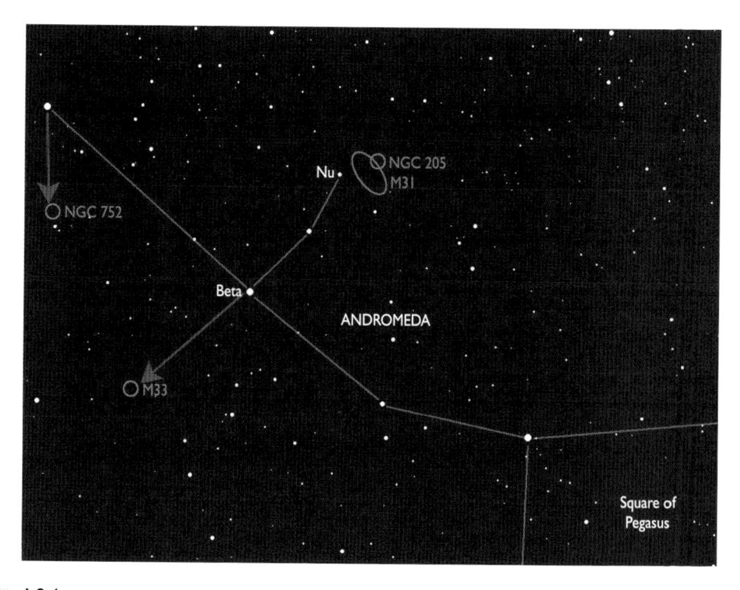

► *A sketch of M31 made by Michael Hezzlewood using 15 × 70 binoculars. To the right is NGC 205, and to the lower left of M31's nucleus is another companion galaxy, M32.*

skies. Also try to visualize the immensity of space. The Milky Way spreads around us, but here we are seeing through it to our twin, and however impressive it may be in binoculars, it is really tiny in our sky.

If you have good, dark conditions you should be able to make out a smaller and fainter object close to the main galaxy. This is its companion galaxy NGC 205, also sometimes called M110. It is larger and brighter than some other binocular galaxies, such as M51, yet it is often overlooked, and is sometimes mistaken for a comet by the unwary!

More nearby galaxies—NGC 253 and NGC 55

The constellation of Sculptor is noted for its edge-on galaxies, NGC 253 and NGC 55. The first is easy to find, lying a bit more than 7° south of the second-magnitude star Beta Ceti, Diphda, the only bright star in the area. A couple of triangles of sixth-magnitude stars help you

► *A photograph of M31 taken from Kitt Peak, Arizona, shows dark lanes which can be seen using large binoculars from a good site.*

◄ *The Andromeda Galaxy, M31, and its companions are easily found by star-hopping from the Square of Pegasus. The map also shows the location of two other objects dicussed on page 108.*

▲ *The edge-on galaxies NGC 253 and NGC 55 in the faint constellation of Sculptor are best seen from more southerly latitudes.*

on your way. The spindle shape of NGC 253 is noticeable even in binoculars. Although not a member of the Local Group, it is one of the closer galaxies to us, at about 9 million light years.

A bit farther south, and therefore particularly difficult unless you are well south yourself, is NGC 55. To find it you need first to locate second-magnitude Alpha Phoenicis, which is roughly midway between Fomalhaut (Alpha Piscis Austrini) and Achernar (Alpha Eridani). It has a fourth-magnitude star, Kappa Phoenicis, just to its south. Go from Kappa to Alpha, turn right by 45° and NGC 55 is three times as far away as the distance between Kappa and Alpha. It is another spindle, of similar size to NGC 253 but somewhat harder to see because of a lower surface brightness. The discovery in 2006 of Cepheid stars in NGC 55 has provided a distance of only 6.2 million light years, making it one of the closest galaxies to the Local Group.

Southern hemisphere only

Small Magellanic Cloud

Though we think of galaxies as giant, spiral, and distant, the Magellanic Clouds are more typical of most galaxies—small and irregular. The

Large Magellanic Cloud (LMC), in Dorado, is described in the text for January, though the two are usually visible together, being circumpolar from more southerly latitudes. Even in October, you need to be south of the equator to see the smaller cloud at all well.

The Small Magellanic Cloud (SMC), like its larger cousin, was so named in honor of the Portuguese navigator Ferdinand Magellan, who saw both clouds on his journey south in 1519. It is both smaller and fainter than the LMC, but also appears oval in shape and is mottled as a result of nebulae and clusters within it. One such object near its edge, however, is actually a much closer globular cluster, NGC 362.

▲ *Chris Picking's sketch of the SMC (upper object) and 47 Tucanae, seen using 10 × 50 binoculars from near Wellington, New Zealand.*

47 Tucanae—second best?

Having found the SMC, you get a bonus object—the great globular cluster 47 Tucanae, which is only 2½° to the west and therefore within the same binocular field of view. This is second only to Omega Centauri in size and brightness. Unlike most other globulars, which are admittedly somewhat underwhelming, 47 Tucanae really does look like a globular cluster of stars, though you need a telescope to be able to see individual stars within it. Though 47 Tucanae may not be the brightest, in terms of location it makes a great

What's a light year?

A light year is a measure of distance, not time. It is the distance light travels in a year, equivalent to 10 trillion km (c.6 trillion mi). Here, a billion means 1000 million, and a trillion is a million million, or 10^{12}. It really makes no sense to use everyday units like miles or kilometers where such huge distances are concerned—who can relate them to any distance they normally travel? The speed of light is just under 300,000 km/sec (about 186,000 mi/sec).

Typical distances are just over four light years to the nearest star to the Sun, tens to hundreds of light years to most of the stars we can see in the sky with the naked eye, thousands of light years to objects within the Galaxy, and a hundred thousand light years or so for the diameter of the Galaxy. Other galaxies are millions of light years away, and the most distant observable objects are billions of light years away.

sight together with the SMC, but it is much closer at about 15,000 light years compared with about 210,000 light years for the SMC.

The November sky

Most of the action in November skies is in the northern part of the sky, because that is where the Milky Way runs at this time of year. The constellation of Cassiopeia is here, its prominent W shape being easy to spot. This is a northern circumpolar constellation as seen from North America and Europe, though it is too far north to be visible from most of the populated parts of the southern hemisphere. To its east is Perseus. South of Cassiopeia lie Andromeda and the three-star constellation of Aries, but south of that are just rather faint stars in Cetus until you get to the brilliant Achernar, marking the southern end of a mostly faint winding celestial river called Eridanus.

The Pinwheel Galaxy, M33

The famous Andromeda Galaxy was described on page 104. Not far from Andromeda is the small constellation of Triangulum, which contains another member of the Local Group, M33, otherwise known as the Pinwheel Galaxy. Find it by looking back at your route to M31. Locate Beta Andromedae and Nu Andromedae again, and look on the other side of Beta from Nu. The Pinwheel is a much fainter galaxy, and unlike M31 it is face on to us, so it appears less condensed. As a result, you need better skies to find it than for M31, but even so it is visible under quite ordinary conditions on a good night. If you are used to galaxy-hunting with binoculars you may be searching for something quite small, but M33 is over ½° across in dark skies.

Open cluster NGC 752

While you are in the area, drop in on the attractive open cluster NGC 752. It lies just about 5° south of the bright star Gamma Andromedae, and has stars of a range of brightnesses scattered over nearly 1° of sky. The double star 56 Andromedae just to the south of the cluster is a good guide to its position, but its two stars are unrelated to the cluster or even to each other. A map of the area is on page 104.

The beautiful Double Cluster

There are some objects that are just beautiful, whatever size of instrument you use to view them, and the two star clusters known as the Double Cluster must rank close to the top of the list. Finding them is usually no problem—they are roughly midway between the main stars

▲ *The Double Cluster in Perseus—a photograph showing stars to magnitude 13, with a field of view of about 5° from left to right.*

of Cassiopeia and Perseus, but if you want more guidance in a poor sky, they are exactly midway between Delta Cassiopeiae and Gamma Persei, as shown on the finder map on page 113. In a half-decent sky you can see them with the naked eye, by averted vision if not directly. The clusters lie within the borders of Perseus.

They are by no means the brightest clusters in the sky, nor the richest, though they do contain stars from magnitude 6 downward. There are numerous faint stars that help to give these clusters added value even when viewed with binoculars that cannot show you the individual stars. Because of this, the Double Cluster does tend to get wiped out by bad light pollution, but as skies improve, so does the spectacle. The fact that there are two clusters side by side reinforces the sight. The clusters themselves go by the numbers NGC 884 and NGC 869, and they are more than 7000 light years away.

Mira the Wonderful

Most bright variable stars remain easily visible throughout their cycle of changes, but Mira is a rare exception. Sometimes it is there, shining red among the other stars of Cetus, the Whale; at other times there is nothing to be seen. It is a red giant star going through a period of

▲ *This map of the stars around Mira has a field of view of 7°. The magnitudes are taken from a chart produced by the American Association of Variable Star Observers.*

▼ *Mira (above center) as seen near maximum brightness at about magnitude 2.4 in the twilight sky in February, 2007. It was the brightest object in this part of the sky.*

instability, and its normal range of brightness is from about magnitude 3 to magnitude 9. It takes 332 days (about 11 months) to go from one maximum to the next. Its peak brightness varies from year to year, and it can sometimes be as bright as second magnitude. It is officially Omicron Ceti, but the name Mira, meaning "Wonderful," was attached to it in the seventeenth century because at the time no other stars were known to behave in such a way.

For binocular users it is interesting to follow it through its cycle of variation, and to locate it when it is just another faint star in Cetus, using the map given here. Magnitudes are given for a few selected stars. There is a magnitude-9.2 star very close to Mira itself.

Northern hemisphere only

Cassiopeia's clusters—NGC 663 and NGC 7789

There are several star clusters in Cassiopeia, but we are singling out just two for binocular observation. Find the first, NGC 663, by bisecting the line between the two eastern stars of the W shape. The cluster lies just to the southeast of this point. It consists of a sprinkling of ninth- and tenth-magnitude stars spread over an area about half the apparent size of the Moon, making it a pretty binocular object for northern observers when the constellation is high in the sky. NGC 7789 is a more subtle cluster and is found by going out about 3° at a right angle from the westernmost stars of the W. It has many faint members and appears as a hazy spot rather than as individual stars. The clusters are shown on the map on page 113.

The December sky

Perseus is on the meridian in December, and is visible from all the northern hemisphere and from the southern hemisphere anywhere north of about Sydney or Cape Town. Though its brightest stars are of second and third magnitude, its shape is quite distinctive, with a line of three stars and another line at an angle to it. Being in the Milky Way, it is rich in stars and is a good area for just plain stargazing.

The star Beta Persei, Algol, is a famous eclipsing binary star (see page 101) which varies in brightness between magnitudes 2.1 and 3.4 over a period of 2 days and 21 hours. This is quite a significant change in brightness, and is obvious to the naked eye. It takes about 10 hours to drop to minimum and back again. Its name means "Demon Star" in Arabic, which suggests that its variability has been known for centuries.

South of Perseus, and visible the world over, is Taurus, with red giant Aldebaran, the V shaped Hyades cluster, and the unmistakable bright cluster of the Pleiades. The skies farther south than this are barren of bright stars.

The Alpha Persei Association

If you find that most star clusters are too faint to be seen from where you live due to light pollution, try this one. Surrounding the star Alpha Persei, or Mirfak, is a host of stars of magnitudes 6 to 8, bright enough to be seen in even bad skies using any binoculars. This area looks just as good at low magnification as any cluster seen through a telescope, and this is no coincidence. This group of stars is referred to as an association, the difference between that and a cluster being that the stars are all moving together through space, rather than necessarily being bound together by gravity. Whatever the distinction, there is a delightful S shape to be seen, which author Robin Scagell refers to as the Switchback.

Cluster M34

This bright cluster is easy to find about 5° west, and a little north, of Algol. Its stars are seventh and eighth magnitude, which makes it an easy object in light polluted skies, though it lacks the fainter stars that improve the spectacle as seen in darker skies.

▲ *A photograph of Mirfak and the Alpha Persei Association, enhanced by the use of a diffraction filter. The "Switchback" of stars is evident.*

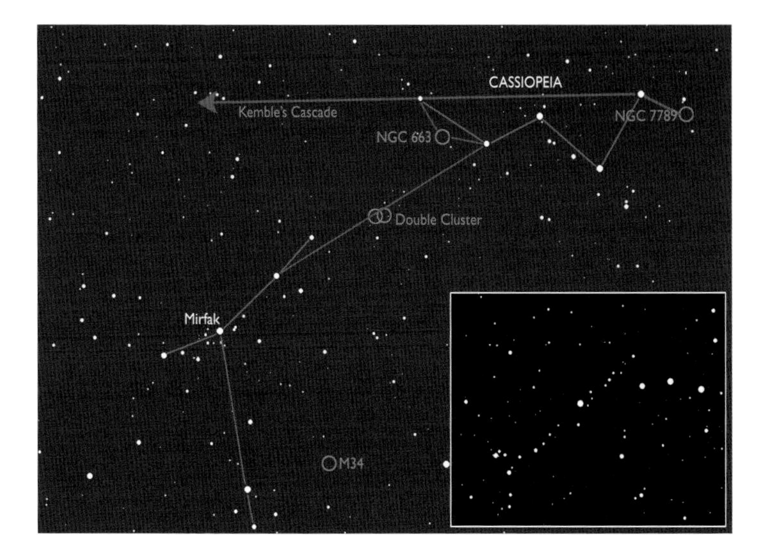

▲ *Perseus and Cassiopeia are excellent signposts for star clusters and Kemble's Cascade (inset, with field of view of about 4°).*

The Pleiades or Seven Sisters

This cluster has to be the showstopper of all clusters. Bright and compact, it is visible from virtually everywhere in the world and grabs the attention even with the naked eye. It is a superb binocular object, as it is almost 2° across and too large to fit fully within the field of view of most telescopes. Everyone returns to this object time and time again for its glittering stars. You can find double stars and star chains, and if you have really dark skies, even some nebulosity.

The name Pleiades (pronounced "Plyadeez") and its popular name of the Seven Sisters refers to its Greek mythological association with the seven daughters of Atlas and Pleione. You would expect there to be seven bright stars. Some people see six with the naked eye, others see nine, and some can see as many as fourteen. You might like to count them before you turn your binoculars on them. Actually, nine have names—but these names include those of the two parents of the seven sisters. The brightest star, Alcyone, is magnitude 2.8, and the other main stars are around magnitude 4. Together, however, they amount to first magnitude, and this is why the cluster can attract the eye even in light polluted areas.

One particularly elusive feature, though often photographed, is the Merope Nebula, a blue reflection nebula that surrounds the southernmost bright star, Merope. Opinions vary, but to be certain of seeing it

▲ *A close-up of the Pleiades, showing stars to about magnitude 13, gives an impression of the appearance of the object in binoculars.*

you need very dark skies and probably large binoculars, though experienced observer Steve Coe of Saguaro Astronomy Club in Phoenix, Arizona, has seen it in 8 × 25 binoculars. Any claims of having seen the Merope Nebula by people observing from average locations must be regarded with the suspicion that there was dew on the objectives!

The illustrious Hyades

The Hyades cluster (pronounced "Hyadeez") is almost too large to be seen in binoculars: at over 5° across, it more than fills the average field of view. The pattern is very distinctive—a V shape of stars with the bright Aldebaran, the angry eye of Taurus the Bull, at one end. The stars are really more of an association than a cluster, as they are widely spaced out and are moving together on parallel paths. Their closeness to the Sun, at an average of about 150 light years, has made them invaluable to astrophysicists studying the properties of stars. Aldebaran, however, is much closer at 65 light years. The V shape has undoubtedly led to the ancient name of the constellation—there is even the suggestion that a cave painting at Lascaux, France, shows the Bull with its V-shaped head, horns, and the Pleiades nearby.

▲ *A long-exposure photograph reveals the reflection nebula surrounding the Pleiades. The Merope Nebula is the southernmost part of the nebula.*

Northern hemisphere only

Kemble's lovely Cascade

Most of the objects we find with binoculars are clusters or other tightly knit objects, but one unusual but attractive sight is actually a straight line of stars, known as Kemble's Cascade after the Franciscan friar Father Lucian Kemble (1922–1999) who first drew attention to it. Kemble observed it with a telescope, but it makes a good binocular object as well.

The Cascade is in the constellation of Camelopardalis, the Giraffe, which lacks bright stars although it is on the outskirts of the Milky Way zone and is plentifully endowed with faint stars. One way to locate it is to take a straight line from the two northernmost stars of the W of Cassiopeia and move eastward by the same distance, which brings you very close to the Cascade. If this proves too hit and miss, you will have to star-hop from Perseus with the aid of the finder map shown on page 113.

Most of the stars of the Cascade are seventh or eighth magnitude, so they should be within reach of any binoculars, though you need skies that will show stars to about magnitude 9 to see it well. Father Kemble would follow the line of stars with his telescope until it ended in a cluster, NGC 1502. This is a very compact grouping of stars of magnitudes 9 to 11, so it is unlikely that you will make much of it unless you are using very powerful binoculars.

4 · THE SOLAR SYSTEM

The Solar System is a surprisingly fertile area for study with binoculars. The planets themselves are all too tiny in binoculars to show any detail, but some of Earth's other neighbors are much more visible, and some, such as comets, are ideally suited to binocular viewing.

There are whole books devoted to describing the Solar System, so forgive us if we don't give much detail here about the actual bodies. The term describes the Sun's family of bodies, which were all formed along with the Sun shortly after it condensed from a nebula some 4.6 billion years ago. The Sun itself is basically a ball of mostly hydrogen gas some 1.4 million km (870,000 mi) in diameter—a typical star, really. Deep inside, where the temperature is in excess of 13 million °C (23.4 million °F), hydrogen is converted into helium with the resultant release of energy. This provides virtually all of the heat and light within the Solar System—all the other bodies shine only because they are illuminated by the Sun. The Sun also contains the lion's share of the mass of the Solar System, so everything else is in orbit around it, mostly moving in the same direction which is counterclockwise as seen from above the Sun's north pole. The Sun itself also rotates in this direction once in 27 days as seen from Earth.

The other main bodies of the Solar System are the planets, which divide into the smaller rocky planets such as the Earth, closer in, and the larger giant gas and ice planets farther out. Between the two groups lies the asteroid belt, consisting of chunks of rock that did not collectively have enough mass to form a large planet; and in the outer Solar System orbit numerous icy bodies known as Kuiper Belt objects or comets. Some comets are occasionally deflected toward the inner Solar System, and when the ice is heated by the Sun they produce quantities of gas which forms a haze around the body or in more spectacular cases streams away in a tail that may be millions of miles long.

The planets close to the Sun orbit the quickest, and have very short years—a planet's year being defined as the time it takes to orbit the

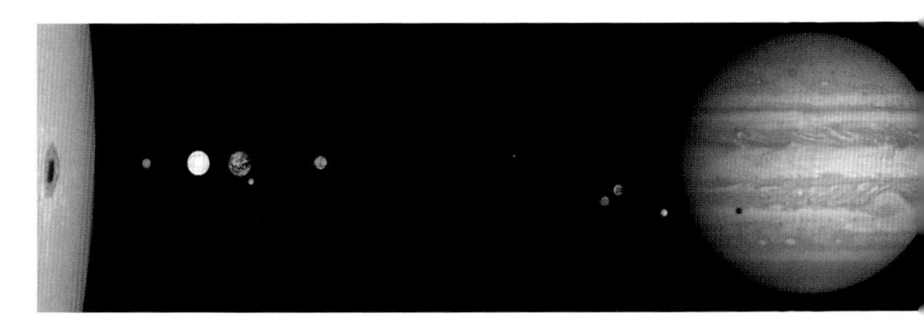

Sun. Farther out, they take longer—Jupiter takes 12 Earth years and Saturn takes nearly 30, for example.

Most of the planets have moons, which are smaller bodies that orbit the planet rather than the Sun. Earth's Moon and the large moons of Jupiter are the only major rocky moons—all the others are mostly small and icy, though they may have rocky cores. Our own Moon is odd in that it is very large compared with its parent planet, and though some of the moons of Jupiter are larger than our own Moon, they are small compared with the planet itself, Jupiter being the largest and most massive of the planets.

All the planets currently defined as such can be seen using binoculars. Pluto lost the battle to be called a planet in 2006, and is now regarded as a dwarf planet, a term which has a somewhat complex definition which need not concern us here. But in any case, Pluto and the other Kuiper Belt bodies are not ordinarily visible in binoculars.

The Sun

Of all the bodies in the Solar System, the Sun is the only one which it can be dangerous to observe. Every child should be taught that it is dangerous to look directly at the Sun, on pain of damaging the eyesight. Fortunately there are ways of observing the Sun safely, as described on page 191, and of course this is a good object for those who don't like losing sleep or object to cold nights.

The Sun appears as a white disc, with the edge or limb appearing slightly more yellow. You may also be able to see sunspots, which are darker areas on the solar disc, where magnetic fields have restricted its light output. They share in the rotation of the Sun, so each one is visible for 10 or 11 days as it moves from one side to the other. As it does so

▼ The Solar System, with the distances between the planets to scale and the planetary diameters 25,000 times the same scale. In order from the Sun: Mercury, Venus, Earth and Moon, Mars, Jupiter, Saturn, Uranus, Neptune. Moons larger than 2500 km (1500 mi) in diameter are also shown.

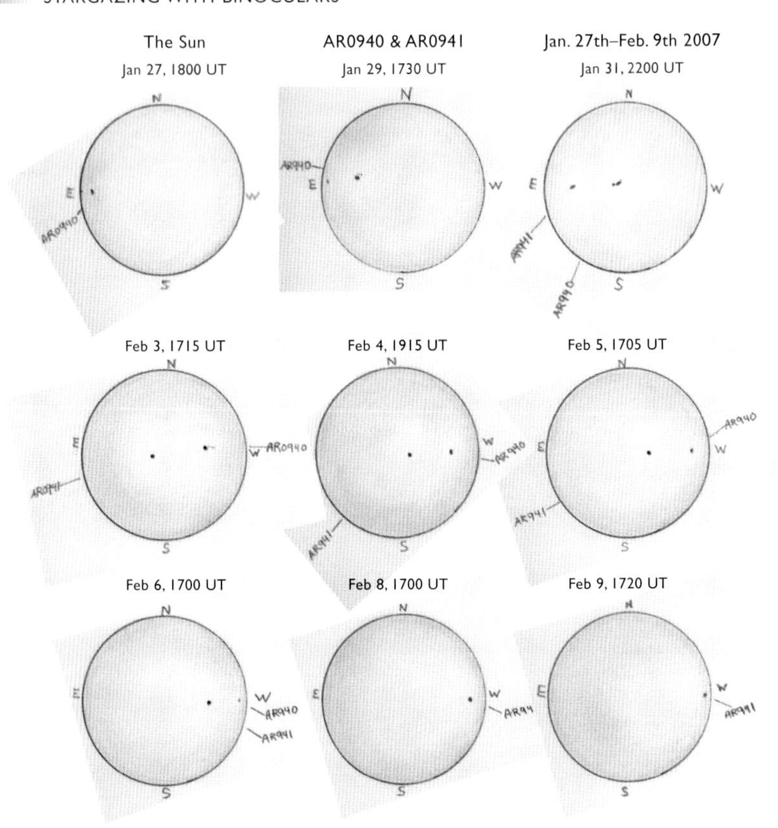

▲ *A sequence of sketches of the Sun made over a period of 13 days using 10 × 50 binoculars and Baader solar filter material by Michael Rosolina in West Virginia, USA. The two spots visible are given Active Region numbers issued by NOAA in the USA, to be found on www.spaceweather.com.*

it may grow or shrink. Sunspots are regularly larger than the Earth itself, and may be the source of particles that spread throughout the Solar System and give rise to aurorae—the northern and southern lights—on Earth.

Sunspots are not always present, as they increase and decrease in number on an 11-year cycle. Even with a telescope, there are times when none at all are visible, and the lower magnification of binoculars is less likely to reveal the smaller spots. There was a minimum in 2006 so one is due in 2017, and in the intervening period it is likely that spots will be visible in binoculars for much of the time.

Spots are often seen in pairs, generally aligned roughly east–west across the face of the Sun. One active area, as it is known, can comprise several separate smaller spots. You can keep your own eye on solar activity by counting the number of active areas on the Sun's surface, and follow their evolution as the Sun rotates. Using larger binoculars, you may also see lighter areas close to the limb, known as faculae, which are clouds of hot gas hanging higher in the Sun's atmosphere that are particularly obvious near the limb.

You may be disappointed not to see the great flames leaping from the edge of the Sun that you see in photographs. These might indeed be present, but they are too faint to be seen without special equipment. These days you can buy telescopes with filters that will reveal these, and some enterprising people have put two of these telescopes side by side to produce effective solar binoculars. But even a single telescope of this sort costs about the same as a large-screen television, so this is for really keen observers.

The Moon

For many astronomers, both amateur and professional, the "dark of the Moon" is the time for real astronomy—seeing those faint fuzzy and elusive objects that we have heard so much about. But the time when the Moon is in the sky need not be wasted by the binocular observer—the Moon itself is always worth a look.

Admittedly, binoculars will not give you the close-up views that turn the Moon into a landscape that some people spend a lifetime studying, but you can see a surprising amount by observing it carefully. Even small binoculars give enough magnification to turn that familiar spotty face into a real world of mountains and craters, and you can learn your

Key

1 Oceanus Procellarum
2 Mare Imbrium
3 Mare Humorum
4 Mare Nubium
5 Mare Serenitatis
6 Mare Tranquillitatis
7 Mare Nectaris
8 Mare Fecunditatis
9 Mare Crisium

◀ *The Moon's major seas are easily spotted using binoculars. The same face is always seen, whatever the phase of the Moon.*

way around very effectively, particularly if you can mount your binoculars steadily and examine the surface carefully.

The "Man in the Moon" which we can see with the naked eye is actually a pattern of dark areas known as seas, or in Latin *maria*, pronounced "mah-rear," with the singular being *mare*, pronounced "mah-ray." These are not seas at all, but are composed of dark basaltic rock that upwelled from the interior of the Moon in a molten state many billions of years ago. Even with the naked eye you can tell that many of them have a roughly circular outline, and this is because they resulted from colossal impacts with asteroids at a time when the Solar System was still in a state of turmoil early in its history. The Moon still shows the scars, but the similar impacts which undoubtedly took place on Earth have all been obliterated by plate tectonics (which gives rise to continental drift) and erosion by weathering. The principal seas are labeled in the illustration on page 119. Get to know them to start with, then you have a framework for finding the other craters.

While you can pick out all these seas together at full Moon, this is not the ideal time to study the fine details, for two reasons. The Moon is so bright when full that it can be quite dazzling to view with binoculars, though you will only temporarily ruin your night vision rather than doing any permanent damage. And the details are much easier to see when they are under low illumination, which emphasizes the relief differences. So look when the Moon shows a phase other than full.

The phases are the result of the Sun's changing angle of illumination on the Moon as the latter orbits the Earth. The cycle begins with the crescent Moon in the western evening sky; it then gets fatter evening by evening until a few days later it is at first quarter (that is, a quarter of the way round its orbit), appears as a half Moon and sets a few hours after the Sun. During the following week it gets progressively more illuminated, through what are called the gibbous phases, until at full Moon it is rising more or less as the Sun sets, but opposite it in the sky, so it is at its highest at midnight. The shadow then starts to cover the areas that were initially a crescent, and the Moon rises later and later, until at last quarter it looks like a half Moon again. During this period the evening sky is increasingly Moon-free, and fewer people get to see its appearance at this phase because it is usually in the early morning.

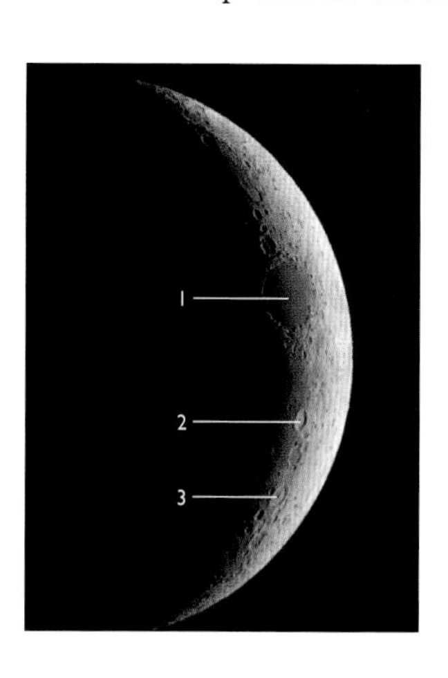

▲ *1: Mare Crisium*
2: Langrenus
3: Petavius

Finally, the waning crescent Moon is visible just before sunrise and there is a period of about a week during which there is no Moon in the sky.

Hundreds of the Moon's craters are named, mostly after long-dead astronomers and scientists. The illustrations of the phases show some of the craters that are most easily identified using binoculars. As you follow the Moon through its phases, you will see that even though a crater might be very obvious when it is on or near the shadow line (the terminator), at other times when the Sun is at a different angle it may be virtually invisible. Even with binoculars you can follow the slow progress of the terminator across the surface. If you examine an area closely every 30 minutes or so, you will notice obvious changes as new peaks catch the sunlight and gleam like jewels in the inky blackness.

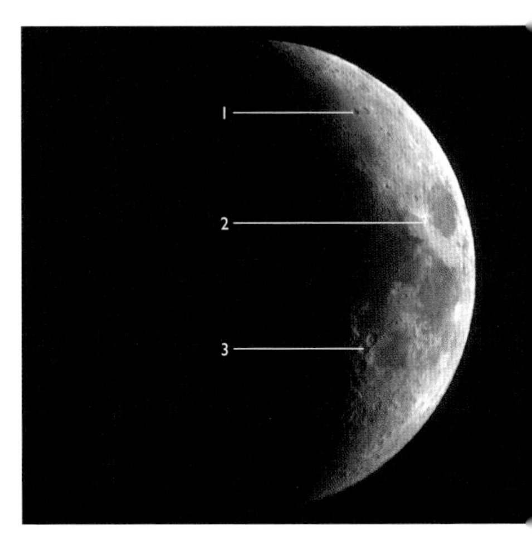

▲ *1:Hercules, Atlas*
2: Palus Somni, Proclus
3:Theophilus, Cyrillus, Catharina

One aspect of the Moon's surface is so familiar to us that we never give it a second thought—the fact that the features on it are always the same, and it does not rotate. This is because it has what is known as captured rotation—it does rotate on its axis, but in the same time that it takes to orbit the Earth once. A number of the Solar System's moons do this with respect to their parent planet—we are not especially favored. Actually, a slight wobble known as libration means that the angle at which we see the visible face does alter slightly over the month. This is particularly obvious if you look at the distance between a feature such as Mare Crisium and the nearest limb.

Also bear in mind that the far side of the Moon is not necessarily dark. Sometimes, as at around new Moon, which is theoretically the moment when it is between us and the Sun, it is the near side that is dark while the far side is fully illuminated. So the popular notion that the other side of the Moon is perpetually dark is wrong.

The Mare Crisium is very easy to spot on the young crescent Moon. It measures 420 × 550 km (260 × 340 mi), but curiously the long axis is east–west, rather than north–south as it appears from Earth. To its north is Cleomedes. Two major craters are easily visible when the Mare Crisium is half in shadow—Langrenus, 133 km (83 mi) across, and

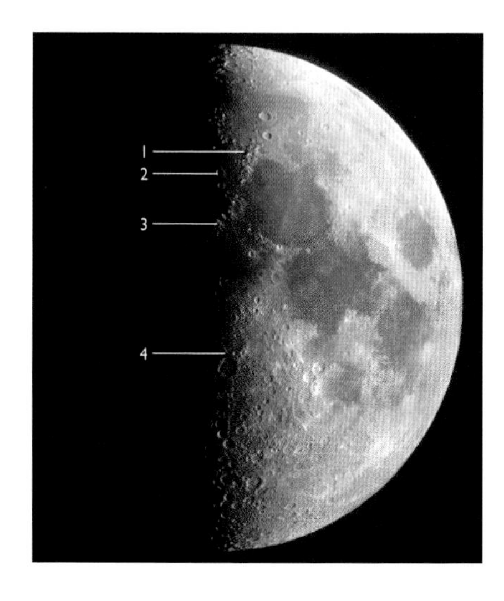

◀ *1: Caucasus Mountains; 2: Aristillus 3: Apennine Mountains; 4: Hipparchus, Albategnius*

Petavius, farther south, at 177 km (110 mi) across. This is known as the eastern limb of the Moon, even though it is westernmost in our sky. Since astronauts walked on its surface, the Moon's cardinal points have been accepted as being those that would apply if you were standing on its surface.

The curious diamond-shaped gray area to the west of the Mare Crisium is Palus Somni. It is actually formed by the edges of two prominent light rays from a crater named Proclus close to the edge of the mare. Many lunar craters have such rays, which are the result of the material that was thrown out of the crater when it was formed by impact. After a few hundred million years the rays become darkened by the effects of solar particles, so we know that any ray craters are younger than most. Atlas and Hercules make an easily spotted pair of craters in an otherwise fairly featureless area. To the south, the trio of Theophilus, Cyrillus, and Catharina catches the eye on the shores of the Mare Nectaris.

▼ *1: Plato; 2: Sinus Iridum; 3: Copernicus 4: Bullialdus; 5: Tycho*

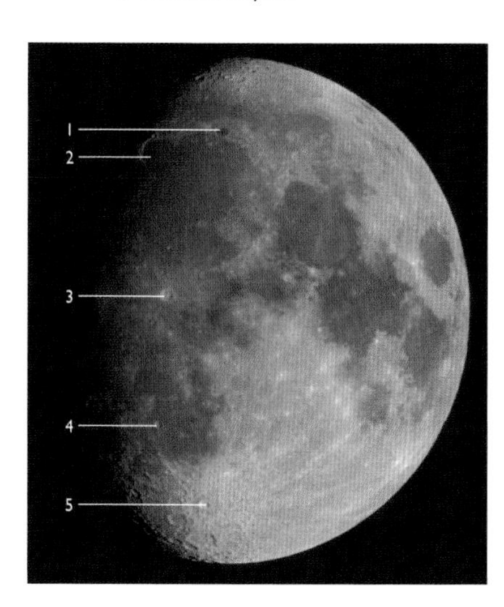

At half Moon, the lunar highlands come into view, and individual craters can be very hard to pick out. Some of these features are very battered by more recent impacts, though here recent is a relative term since we are talking about a landscape that is billions of years old. Hipparchus and Albategnius are obvious in this view, though when they are farther from the terminator they all but disappear. Farther north the lunar Apennines and Caucasus mountains are coming into view,

► *At full Moon, many craters are hard to spot because of the high illumination, but the difference in brightness between the highland areas and the seas is obvious.*
1: Aristarchus
2: Kepler
3: Grimaldi
4: Schickard
5: Mare Frigoris
6: Mare Vaporum
7: Sinus Medii

separating Mare Serenitatis from Mare Imbrium. Aristillus on the floor of Mare Imbrium is 55 km (34 mi) across.

A few days later the Moon is at gibbous phase. On the terminator now, on the shores of Mare Imbrium, is the beautiful bay Sinus Iridum, 250 km (150 mi) across. To the north of Imbrium the 100 km (60 mi) dark-floored walled plain Plato is very distinctive. Several ray craters dot the Oceanus Procellarum, the most obvious of which is Copernicus, regarded as a recent crater at 800 million years old. But the most dramatic ray crater of all is farther south, on the highlands—Tycho, whose rays cover a huge area of the surface.

Tycho now dominates the southern half of the Moon, with the rays from Copernicus and Kepler being very evident even though with binoculars the craters have merged into the background. However, another even brighter feature is now gleaming in the western Oceanus Procellarum—the brilliant crater Aristarchus. This is the brightest crater on the Moon. Notice how the eastern part of the Moon now lacks much detail other than bright and dark spots.

At full, the only detail on the Moon comes from brightness differences. The ray craters stand out, while the dark, walled plain Grimaldi is now very noticeable near the western limb. Usually there is a little shadow to be seen, even when the Moon is theoretically full, because the orbit of the Moon is tilted at an angle to the ecliptic, the path of the Earth round the Sun. This means that there is almost always some shadow somewhere on the Moon.

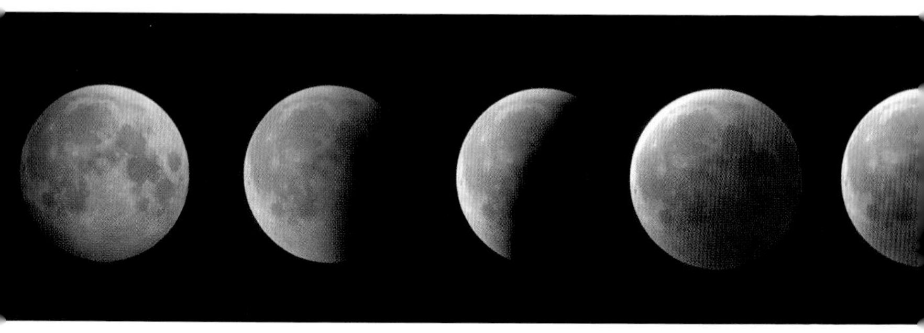

Lunar eclipses

A lunar eclipse takes place when the full Moon is exactly opposite the Sun, so it passes through the Earth's shadow. Because of the tilt of the two orbits, the Moon usually misses it, but every six months or so there is a chance of an eclipse, though it may not be visible from where you happen to be. It is not always a total eclipse, as the Moon may not necessarily pass through the center of the shadow. Partial lunar eclipses are those where some part of the Moon remains in sunlight, so it always looks as if it has a curved bite taken from it. A penumbral eclipse is where the Moon passes through the very edge of the Earth's shadow, and is hardly dimmed at all.

Binoculars really are the ideal instruments for observing a lunar eclipse. The sight of what should be a full Moon hanging in the sky, but as a pale and ghostly body instead, in the case of a total eclipse, is dramatic and beautiful. Stars are visible surrounding it, and you can see stars that would normally be lost in its glare winking out or reappearing as the Moon passes in front of them. The total phase can last for well over an hour, giving the opportunity to savor the sight and even get in some deep sky observing as well. During the partial phases, you can see so much more of the shadow stealing slowly across the lunar surface than with the naked eye alone. In the case of a total eclipse, binoculars bring out its color more strongly than you can see with the naked eye.

No two total eclipses are the same. For one thing, the path that the Moon takes through the Earth's shadow varies from being central to close to one edge. The color of the shadow varies from place to place. Near the edge it is usually bluish, while in the center it can be red, orange, yellow, or coppery. These colors are caused by the sunlight that filters through the Earth's atmosphere, for just the same reason as the Sun goes red at sunset. If the edge of the Earth happens to be particularly clear, more red light will get through; if it is cloudy, the eclipse may be very dark. Sometimes an eclipse is so light that the

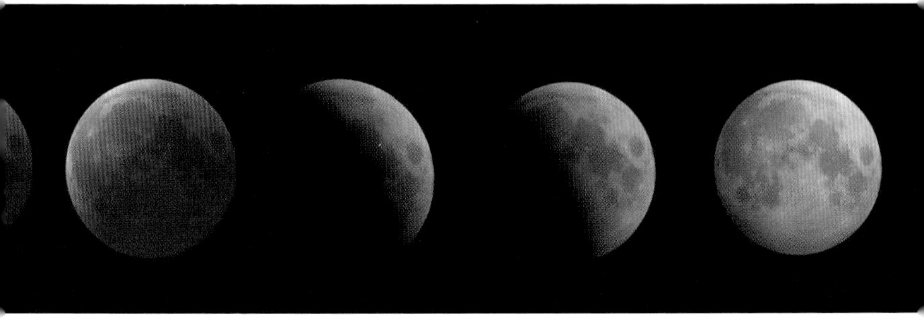

▲ *A sequence of the total lunar eclipse of March 3, 2007. It begins with the Moon already within the outer shadow of the Earth, known as the penumbra. Gradually the main shadow encroaches until it completely covers the Moon. The three central images have a greater exposure time to bring out the color, but to the eye the Moon appeared very dark and the sky was virtually as dark as on a moonless night. The complete sequence took five hours to unfold.*

casual observer may not notice anything very unusual—though this is partly because these days people are not very aware of the Moon, and don't realize that it is normally only red when low in the sky. At other times, the Moon is so dark that it is hardly visible at all, and appears as just a strange sliver of light. A major volcanic eruption on Earth can also have the effect of making our atmosphere darker, leading to a dark eclipse.

There is another trick that you can use with binoculars. There is some interest in knowing the brightness of the eclipse. By inverting your binoculars, so that the Moon is reduced in size to that of a star, you can compare its brightness to any nearby bright stars. This is hard to estimate in any other way. Alternatively, you can measure the estimate with the naked eye on a scale devised by the French astronomer André-Louis Danjon, which gives it a value, L, as follows:

L = 0 Very dark eclipse. Moon almost invisible, especially at mid-totality.
L = 1 Dark eclipse, gray or brownish in coloration. Details distinguishable only with difficulty.
L = 2 Deep red or rust-colored eclipse. Very dark central shadow, while outer edge of umbra is relatively bright.
L = 3 Brick-red eclipse. Umbral shadow usually has a bright or yellow rim.
L = 4 Very bright copper-red or orange eclipse. Umbral shadow has a bluish, very bright rim.

Solar eclipses

These occur at the opposite point in the Moon's orbit from lunar eclipses, that is, at new Moon, and are caused by the Moon getting between the Earth and the Sun. While a lunar eclipse can be seen from anywhere on Earth that the Moon happens to be visible during the event, a solar eclipse can only be seen from somewhere within the shadow of the Moon as it falls on the Earth. If you are within the edge of the shadow, you see a partial eclipse of the Sun, so it appears to have a bite out of it. But if you are lucky enough to be centrally located on the eclipse path, you see the Moon completely covering the Sun. As the two bodies, by chance, are roughly the same apparent size, the Sun's outer atmosphere becomes visible for a few minutes. Total solar

▼ *Two views of the same total solar eclipse. To the left, a long exposure that shows the outer streamers of the corona, the Sun's extended atmosphere. Inset to the right, a short exposure showing the inner corona and several small prominences. While photographs can capture only a limited brightness range, the binocular user can see all these details simultaneously.*

eclipses are among the most exciting and awe-inspiring spectacles that the heavens have to offer.

People may travel to the ends of the Earth to be within this center line, and often they can take only binoculars with them. Again, these are the ideal instruments for viewing the event. High-powered telescopes are unnecessary, as what you want to see is the surroundings of the Sun rather than any fine details.

However, there is a big caution. During the partial phases of a solar eclipse, the Sun's surface is still literally blindingly bright, and you must take the same precautions as when viewing the Sun at any other time (see page 191). The only time when you can view the Sun directly with binoculars is during totality itself. Even during an annular eclipse, which is when the Moon's diameter is slightly smaller than that of the Sun, you can't use binoculars directly even though only a thin ring of the Sun's bright surface is visible. This is still enough to burn a ring-shaped scar on your retina.

So during a total eclipse you must wait until the last vestiges of bright light have disappeared, and then you can view the Sun in safety using your binoculars. In this health-and-safety conscious age, many authorities now try to persuade you that watching the eclipse on television is the only safe option. This is vastly overprotective as there are countless human activities far more dangerous than viewing a total eclipse with binoculars. But it is imperative that you stop watching as soon as the Sun starts to reappear.

When you do look, you will see the delicate silvery streamers of the Sun's corona, which are caused by material constantly streaming off the Sun. You will, with any luck, see some prominences, which are red clouds of hydrogen that hang over active areas on the solar surface and which can extend for thousands of miles into space. You may see bright stars or planets that happen to be close to the Sun but which are hidden by the Sun's glare. And you may even spot a comet that is making a close dash past the Sun, and which is also usually impossible to see from the Earth's surface.

These days, we know what the appearance of the sky will be in advance, and spacecraft such as SOHO (Solar and Heliospheric Observatory) can reveal prominences and any comets that have put in a brief showing but which were not previously seen, so there are not likely to be any surprises in the view. But the sheer excitement of seeing the Moon's shadow looming over you, the sky going dark during daylight, a hot day turning cool as the Sun leaves the landscape, and the amazing sight of the Sun being turned into a pearly ring is enough to cause people to travel across the world for an event lasting just a few minutes.

Mercury

You have to be quick to see Mercury. It is only visible for a few days at a time in the twilight sky, either before sunrise or after sunset, and then for a fairly short time. If you want to know when to look, the table opposite shows a list of dates of visibility. When the elongation is at a maximum you should look in the general direction indicated. However, there is another factor to take into account, which is the angle of the ecliptic to the horizon. Mercury is best seen when the ecliptic makes a steep angle to the horizon, which occurs in spring evenings or autumn mornings whichever hemisphere you are in, though the months in each case are different. At the wrong time of year, Mercury hugs the horizon to such an extent that to all intents and purposes it is invisible.

Even with a telescope. Mercury presents a very small disc so with binoculars there is no hope. All you will see is a starlike object in the twilight sky. But many people say they have never seen Mercury, so at least you can say you have spotted it.

▼ *Juan Carlos Casado took this sequence of the visibility of Mercury over 22 days from a clear site in Spain. The earliest images are to the right.*

MERCURY								
Max. elongation		**Visibility**	**Separation**	**Max. elongation**		**Visibility**	**Separation**
2008	May 14	Evening	22°	2014	September 22	Evening	26°
2008	July 2	Morning	22°	2014	November 2	Morning	19°
2008	September 11	Evening	27°	2015	January 15	Evening	19°
2008	October 22	Morning	18°	2015	February 25	Morning	27°
2009	January 5	Evening	19°	2015	May 7	Evening	21°
2009	February 14	Morning	26°	2015	June 25	Morning	22°
2009	April 26	Evening	20°	2015	September 4	Evening	27°
2009	June 13	Morning	23°	2015	October 16	Morning	18°
2009	August 25	Evening	27°	2015	December 29	Evening	20°
2009	October 6	Morning	18°	2016	February 7	Morning	26°
2009	December 19	Evening	20°	2016	April 19	Evening	20°
2010	January 27	Morning	25°	2016	June 5	Morning	24°
2010	April 9	Evening	19°	2016	August 17	Evening	27°
2010	May 26	Morning	25°	2016	September 29	Morning	18°
2010	August 7	Evening	27°	2016	December 11	Evening	21°
2010	September 20	Morning	18°	2017	January 19	Morning	24°
2010	December 2	Evening	21°	2017	April 1	Evening	19°
2011	January 10	Morning	23°	2017	May 18	Morning	26°
2011	March 23	Evening	19°	2017	July 30	Evening	27°
2011	May 8	Morning	27°	2017	September 12	Morning	18°
2011	July 20	Evening	27°	2017	November 24	Evening	22°
2011	September 3	Morning	18°	2018	January 2	Morning	23°
2011	November 14	Evening	23°	2018	March 16	Evening	18°
2011	December 23	Morning	22°	2018	April 30	Morning	27°
2012	March 5	Evening	18°	2018	July 12	Evening	26°
2012	April 19	Morning	27°	2018	August 27	Morning	18°
2012	July 1	Evening	26°	2018	November 7	Evening	23°
2012	August 17	Morning	19°	2018	December 15	Morning	21°
2012	October 27	Evening	24°	2019	February 27	Evening	18°
2012	December 5	Morning	21°	2019	April 12	Morning	28°
2013	February 17	Evening	18°	2019	June 24	Evening	25°
2013	April 1	Morning	28°	2019	August 10	Morning	19°
2013	June 13	Evening	24°	2019	October 20	Evening	25°
2013	July 30	Morning	20°	2019	November 28	Morning	20°
2013	October 9	Evening	25°	2020	February 11	Evening	18°
2013	November 18	Morning	19°	2020	March 24	Morning	28°
2014	January 31	Evening	18°	2020	June 5	Evening	24°
2014	March 14	Morning	28°	2020	July 23	Morning	20°
2014	May 25	Evening	23°	2020	October 2	Evening	26°
2014	July 13	Morning	21°	2020	November 11	Morning	19°

Venus

If Mercury is shy, Venus can dominate the twilight sky for months at a time. Its movements are similar to those of Mercury in that it always remains fairly close to the Sun, and either follows the Sun down after sunset or rises before it in the morning sky. But it can get much farther from the Sun than Mercury and can sometimes be seen in a dark sky. It is also a brilliant object, and is the brightest of all the planets. A table of visibility is given below and the same rules apply about the angle of the ecliptic. At an unfavorable appearance, it may scarcely be visible, particularly if you live at high latitudes, including Canada. Venus always moves to a maximum of 45° to 47° from the Sun, so this figure is not so important in its visibility.

However, unlike Mercury, there is the chance of seeing something more than a point of light with binoculars. As it moves around the Sun, Venus displays phases like those of the Moon. When on the far side of the Sun it is a tiny disc, fairly gibbous, while at its greatest elongation from the Sun it appears at half phase and is somewhat larger. When near Earth it shows a crescent phase, and in fact is then closer than any other planet. During this phase in particular it is large enough to be visible with binoculars. Keen-eyed observers may also detect the planet's half phase if they are equipped with binoculars with a magnification of 10× or higher. The crescent phase is vis-

VENUS		
Max. elongation		**Visibility**
2009	January 15	Evening
2009	June 6	Morning
2010	August 20	Evening
2011	January 9	Morning
2012	March 27	Evening
2012	August 15	Morning
2013	November 1	Evening
2014	March 23	Morning
2015	June 7	Evening
2015	October 26	Morning
2017	January 13	Evening
2017	June 4	Morning
2018	August 17	Evening
2019	January 6	Morning
2020	March 25	Evening
2020	August 13	Morning

▲ *Venus is often referred to as the Evening or Morning Star, and regularly makes a delightful sight when teamed up with a crescent Moon.*

▲ *Venus as it appears in the daytime as a crescent (center). It looks like the Moon, but is about 40 times smaller, so the more magnification the better.*

ible both toward the end of an evening appearance and at the beginning of a morning appearance.

Unlike Mercury, Venus can be seen in a dark sky, and you might think that this is the best time to look at it. However, it is so dazzling against the dark sky that you will see very little, so it is best to try and pick it up in as bright twilight as you can, when it appears against a blue sky. It is perfectly feasible to spot it during the middle of the day, by estimating its elongation from the Sun and working out where the ecliptic lies, but you must always be sure to keep the Sun hidden behind something so as to avoid looking at it directly.

A computer planetarium program will show you when the Moon is close to Venus, and this can help when trying to locate it in daylight.

Mars

From the binocular observer's point of view, Mars is only a point of light. But like the other planets, the fact that it is not a starlike object and has some size means that it is far less likely to twinkle than does a genuine star. Twinkling is due to the distortion of the incoming light beams by turbulence in our atmosphere, so it is strongest for the tiny points of light of stars, and minimal for the planets, at least when they are high in the sky. The lack of twinkling and its salmon-pink color are often enough to distinguish Mars from the stars, particularly when it is near opposition—that is, when it is opposite the Sun in the sky and appears at its brightest.

The positions given opposite tell you in which constellation to find Mars on the date of opposition, and its magnitude. But the planet will not always remain in that constellation. It begins its period of visibility in the early morning sky, rising just before the Sun, then rises earlier and earlier in the morning as it comes closer to Earth, getting brighter all the time. At opposition the planet is on your meridian at midnight, and for a month or two on either side of that date it remains in or near the constellation shown. Following this, it moves more into the evening

▼ *Mars as seen in the sky (left), compared with Spica in Virgo (center), appears distinctly reddish particularly when viewed with binoculars.*

sky and is observable at a more social hour, but becomes fainter until after several months it is lost in the evening twilight. Mars returns to opposition every two years, though for about six months of that time is virtually unobservable, being too close to the Sun for practical observation.

All you will see using binoculars is a bright dot in the sky, but you can appreciate its strong color more with binoculars than with the naked eye. It may be called the red planet, but it is hardly bright red and is much more muted in its hue—though to the ancients its reddish tinge reminded them of blood, so it got a reputation as being the planet of war.

The change in magnitude at opposition tells you whether it is a close or a distant approach. Mars has a very elliptical orbit, so its closest distance can vary considerably. In 2012, for example, it will come no closer than 100 million km (60 million mi), while in 2018 it will be within 58 million km (36 million mi) and its disc over 25 arc seconds across, so on this occasion binoculars will show more than a point of light, though no surface details.

MARS			
Opposition		**Constellation**	**Magnitude**
2010	January 30	Cancer	–1.3
2012	March 4	Leo	–1.2
2014	April 9	Virgo	–1.5
2016	May 22	Scorpius	–2.0
2018	July 27	Capricornus	–2.8
2020	October 14	Pisces	–2.6

▲ *Jupiter as it appears at about 15× magnification. No detail is visible, but the satellites are easily spotted.*

Jupiter

Like Venus, Jupiter is a planet that can show a visible disc through binoculars, especially those with higher power. Its four bright moons are also visible, and you can even see them moving over a period of time.

Jupiter remains in one constellation for a whole year at a time, on average, so there is little doubt about its identity. It goes through

JUPITER		
Opposition		**Constellation**
2008	July 9	Sagittarius
2009	August 15	Capricornus
2010	September 21	Pisces
2011	October 29	Aries
2012	December 3	Taurus
2014	January 6	Gemini
2015	February 7	Cancer
2016	March 8	Leo
2017	April 8	Virgo
2018	May 9	Libra
2019	June 11	Ophiuchus
2020	July 14	Sagittarius

the same period of visibility as Mars, from its first appearance in the morning sky to opposition, then into the evening twilight, but its brightness and indeed size variations are much less marked because it is farther out in the Solar System and the change in distance as Earth moves round the Sun is proportionately less. Jupiter is the brightest planet after Venus, and is a pure white color.

Through even 7× binoculars you can see that Jupiter is more than a starlike point, but you need a magnification of 20× or more to see the slightly flattened disc and maybe even a dark belt or two. But its four major moons, Io, Europa, Ganymede, and Callisto, are visible like little beads on a string, close to the planet. They are quite bright—fifth magnitude—but the glare from Jupiter can hide them to some extent. Look from night to night and you will see that their positions have changed, and you may even spot a difference over the course of an evening.

It was this movement around the planet that made Galileo realize in the seventeenth century that the Earth was not the center of all movement in the Solar System, as doctrine had it at the time, and that maybe there was something to the odd ideas of Copernicus after all. Your binoculars certainly give a better view of the planet than Galileo's first telescope, so you can reflect on the importance of what you see.

Saturn

Every small telescope proclaims on its box that it will show you the rings of Saturn. Actually, they are not that hard to spot, though lower-power binoculars will just show you that the planet is slightly elongated. Find it by referring to the table below, in which you will find that it remains in or near the same constellation for three years or so at a time. It takes nearly 30 years to orbit the Sun once, so it is visible at roughly the same season and in the same part of the sky for years on end. Its yellowish tinge is quite noticeable compared with Jupiter, though it is not as bright and can often be mistaken at first for a rather bright star, though its lack of twinkling gives it away.

SATURN

Opposition		Constellation
2009	March 9	Leo
2010	March 22	Virgo
2011	April 4	Virgo
2012	April 16	Virgo
2013	April 28	Libra
2014	May 11	Libra
2015	May 23	Libra
2016	June 3	Ophiuchus
2017	June 15	Ophiuchus
2018	June 28	Ophiuchus
2019	July 10	Sagittarius
2020	July 21	Sagittarius

► *Changes in the tilt of Saturn's rings between 1990 and 2004, as photographed by the Hubble Space Telescope.*

▲ *Jupiter (left) and Uranus in the same part of the sky, Capricornus. Uranus was magnitude 5.7. Stars are shown to magnitude 8.*

As with Jupiter, the view through binoculars is similar, and probably better, to the one that Galileo saw for the first time 400 years ago. But he did not have the advantage of knowing what Saturn actually does look like. With enough magnification, it is possible to persuade yourself that you can actually see the rings, and certainly Saturn is noticeably different from Jupiter or Mars. But if you had no idea what you were seeing, would you imagine that there could be something as amazing as a thin ring completely encircling the planet? Today, every space cartoon shows planets with rings, but in the seventeenth century nothing of the sort existed as a comparison. In fact, all Galileo could say in 1610 was that Saturn had a strange tri-form appearance, seeing the lobes of the rings as possible stationary moons. It was not until 1655 when Christiaan Huygens, using a better telescope, realized that the ring surrounds the planet and nowhere touches it. Huygens also discovered Saturn's largest moon, Titan, which you may see as a eighth-magnitude star close to the planet.

One factor affecting the visibility of the ring system is their tilt as seen from Earth. As Saturn orbits the Sun every 29½ years, the aspect of the rings goes through a cycle from fully open through to edge on and back again so as to be fully open on the other side. When the rings are fully open, as in 2016 and 2033, Saturn appears very much larger through binoculars and the rings are more evident. But in 2009 and 2025 the rings appear exactly edge on, so for some time Saturn appears as just a small disc in binoculars. So thin are the rings that even users of telescopes have difficulty seeing them at these times.

Uranus and Neptune

These two planets were unknown to ancient astronomers because of their faintness. Uranus can be bright enough at magnitude 5.6 to be seen with the naked eye, but it would take a very dedicated ancient astronomer to notice or even care that one apparent star close to the limit of visibility moved position from month to month. Neptune is beyond naked-eye visibility, being magnitude 7.8 at its brightest. Both are visible in binoculars, but the problem is that you really do need a star map that goes fainter than the planet itself, so that you know which it is; and this would occupy many pages. So if you want to find them your best course of action is to use some computer software that will print out a map for you, or use an online resource (see www.stargazing.org.uk for suggestions).

If you do use a map, you will need some care to make sure that you are looking at the right object, as both planets appear essentially starlike in binoculars. Both of them appear noticeably blue in color compared with the other planets, though no more so than many stars. You may need to compare the area with your star map and return on another night to be sure that your suspected planet has moved. A few nights later may be enough if the planet happens to be near to a star, but otherwise you may have to wait a week or two to be sure.

Asteroids

Several asteroids are bright enough to be within binocular range, but the same problems of identification apply as for Uranus

▶ *Vesta, seen two nights apart. The minor planet is at center in the top picture, but has moved to the right in the lower picture.*

and Neptune. The brightest objects are Vesta, which can reach magnitude 5.5, Ceres, the largest, at magnitude 6.7, and Pallas, also at 6.7. The asteroids move considerably faster through the sky than Uranus or Neptune, so a day or two should make all the difference in position.

Very occasionally, a near-Earth object is discovered that races past the Earth and becomes bright enough to be visible with binoculars. One would not like these to approach too close in fact, as they are potentially very dangerous. However, on those occasions when one becomes visible, you may be able to see movement within even a matter of a few minutes. Again, star maps are needed, and at such times they may be published on the websites of astronomy societies and in magazines.

Comets

The sudden appearance of a bright comet is something that every amateur astronomer hopes for. They are few and far between. Comet Hale-Bopp put on a spectacular show in 1997, and before that there was Comet Bennett in 1969 and, briefly, Comet West in 1976. In 2007, Comet McNaught rivaled them all as seen from the southern hemisphere. When a new bright comet shows up, it becomes headline news. Often the period of visibility is short, and restricted to twilight.

There are no bright comets predicted to appear for the foreseeable future, but comets can turn up unexpectedly from the outer Solar

▲ *Comet Linear of 2001 was a fainter comet that was easily visible through binoculars. The two photographs were taken a few minutes apart, and the comet's movement against the starry background is noticeable.*

▲ *Comet Hale-Bopp was a dramatic sight for everyone in early 1997, but such bright comets are rare, and often appear with little warning.*

System at any time. Most of them are rather faint, but every year or two there is one that is bright enough to make a pleasing sight, even if it doesn't make the headlines. All you can do is to subscribe to astronomy society and magazine email alerts to be sure of catching them.

Your binoculars are often ideal for spotting such comets, because frequently they have short tails extending for 1° or more, just right for the field of view of most binoculars. Telescopes may magnify too much to reveal the full extent of the tail, and even the dedicated observers reach for their trusty binoculars at such times. You can sometimes see detail within the tail, as spurts of gas are emitted periodically, causing a brightening.

Comets are notoriously unpredictable, and from time to time what is expected to be a faint comet suddenly flares up and provides a treat. But at other times, what seemed to be a promising object shows little more than a haze around it, looking very much like a globular cluster. More than one observer has mistaken a globular cluster for a comet, and vice versa.

Though they appear to hang in the sky on any particular night, comets generally move fairly quickly through the sky from night to

▲ *Comet Swan (lower left) passed near to the globular cluster M13 (above center) in 2006. This view comparing the two was drawn onto a pre-prepared star map by Jeremy Perez of Flagstaff, Arizona, USA, using 10 × 50 binoculars.*

night. If one chances to move into a dark part of the sky, it can remain visible for weeks as it moves (usually) away from Earth. A comet is often considerably brighter after its closest approach to the Sun (known as perihelion) than beforehand.

Meteors

Most people have seen a shooting star or meteor at some stage, and indeed they are very common, particularly on certain nights. They are caused by tiny particles of dust from comet tails that have spread far from their parent comet, and happen to plunge into the Earth's atmosphere at a speed of many miles a second, burning up as they do so. The whole show is usually over in a fraction of a second, and though it is possible to say when larger numbers than usual will be seen, the exact moment and position of an individual one is a matter of chance. So binoculars are not the ideal instruments for studying meteors, and the usual technique is simply to gaze at as much of the sky as possible for a long period, making notes of the numbers seen.

However, for every bright meteor there are several fainter ones. Although binoculars restrict your field of view considerably compared with the naked eye, the increased numbers mean that seeing a meteor through binoculars is not such a rare thing. So while it is not worth gazing through binoculars in the hopes of seeing a meteor, don't be surprised if one appears as you are watching. If it is a fairly bright one you might see a column of gas expanding away from its path, just for a second or so. Others are much fainter, and look just like a swiftly moving star. However, if you find that you can track the object through the sky as you watch, it is not a meteor but a satellite.

Satellites

Artificial satellites are increasingly common, and are a feature of the sky in the summer months in particular. This is because they are only visible when illuminated by sunlight, and during winter around midnight when the Sun is a long way below the horizon it is less likely that satellites will be illuminated, though some are always high enough to be seen. The very brightest objects, such as the International Space Station, are brilliant and unmistakable, but there are hundreds of fainter ones that can move through the field of view of binoculars. They generally remain visible as they move over many degrees of sky, though quite often a satellite will move into the Earth's shadow and will slowly fade. And of course there is always the chance that you are looking at a high-altitude aircraft, though usually you can see red and green wingtip lights.

The website www.heavens-above.com is an excellent way to find out what bright satellites are visible from your location.

5 · CHOOSING BINOCULARS

It is not until you decide to buy binoculars that you discover that the choice is not straightforward. You may want something not too expensive but good enough for astronomy and general viewing—but start to look around and you will find that there is a very wide choice indeed. So this chapter will tell you about the different choices you have and how to go about testing and buying binoculars.

Binoculars are used by a surprisingly wide range of people. In addition to astronomers, there are bird- and wildlife-watchers, sailors, race-goers, the military, and probably hundreds more specialized users. Each has a certain set of requirements, and in many cases an instrument designed for one purpose might not suit everyone. So it helps to get to know the different types and what they are designed for. To start with, here is a brief outline of what you can do with binoculars of a selection of different sizes.

▲ *A selection of binoculars suitable for astronomy. L–R (front): Nikon 10 × 35; Canon 10 × 30 image stabilized; Bushnell 8 × 42; Pentax 10 × 50; (back): Baigish (Russian) 20 × 60; Revelation 15 × 70.*

8 × 20	Pocket binoculars, helpful for occasional use during daylight. A bit dim for astronomy, but will show lunar craters, reveal numerous stars that are invisible with the naked eye, such as the fainter stars of the Pleiades or Hyades clusters, and will enable you to view some deep sky objects.
7 or 8 × 40	A good, lightweight all-round instrument. At night will show lots of stars in otherwise apparently blank areas, and is useful for picking out a wide range of the brighter deep sky objects.
7 × 50	Often regarded as ideal for astronomy, giving a very bright image with a wide field, though only younger people can use all the light provided. Good for faint but large objects, such as the larger clusters and nebulae, but less useful for picking out galaxies and most planetary nebulae. The wide field is useful for locating objects for subsequent study using a telescope, because it is fairly easy to compare the naked-eye view with the view as seen through the binoculars.
10 × 50	The basic instrument for many astronomers, giving a good compromise between magnification and aperture, though 50 mm binoculars can be a little heavy for prolonged use. Will also give excellent day views, and will reveal many of the smaller deep sky objects that are tricky with the lower-powered 7 × 50s.
12 × 45	Good for stars in particular, and for observing from light polluted areas. Despite the slightly dimmer view compared with 10 × 50s, will still show a wide range of deep sky objects. Less useful for locating objects, as the narrower field of view means that you are less able to know exactly which bit of the sky you are looking at compared with the naked-eye view.
15 × 70	The dedicated amateur astronomer's binoculars—definitely too large and heavy for extended hand-held use, but favored for giving detailed views of the sky and deep sky objects, particularly if mounted on a tripod.

This is by no means a full list of the range available, but it gives you an idea of what can be done with each size. In addition, image-stabilized binoculars tend to show you more for their size than conventional models.

Types of binoculars

No matter what their intended use, there are two basic types of modern instrument. The traditional binoculars have staggered barrels on each half, making the front lenses considerably wider apart than the distance between your eyes. A more modern design has straight-though barrels, which makes them considerably more compact. Actually, the very first field glasses also had this appearance, but there is a major difference between those and the modern variety. To see why there is a difference we need to know what is inside the instruments.

If you bear in mind that binoculars are really two telescopes side by side, think about the traditional telescope. Of necessity it has quite a long barrel—much more unwieldy than binoculars. There is a main lens, or objective, at the front and an eyepiece at the back that you look through. The principle is easy to follow. At some point you have probably played with a magnifying glass, and have found that you can form a small upsidedown and back-to-front image of a window on the opposite wall. You could look at the details in this image using a second, more powerful magnifying glass. In a practical telescope, instead of looking at the image projected onto a white screen, you look at it from the back using your second magnifying glass, which we call the eyepiece. This also gives a much brighter view, though it is still upside down.

The distance from the main magnifying glass to the image on the wall is called its focal length (actually, this is only true if the window is

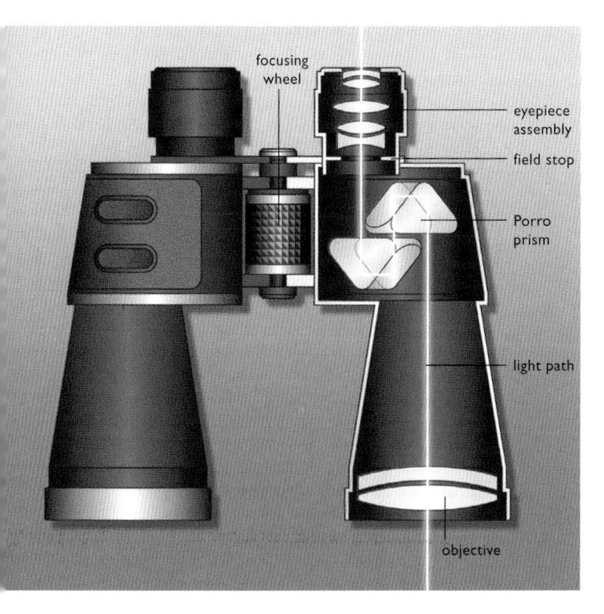

◀ *Porro prism binoculars use two right-angle prisms to fold and invert the light beam, making the instrument much more compact than the equivalent telescope.*

infinitely distant, but the idea is what matters). To make a telescope with any reasonable quality you need an objective with a fairly long focal length—at least five times its diameter, and often longer. So with an objective 50 mm (2 in) in diameter, the focal length would be 250 mm (10 in). Allow room for the eyepiece on the other side of that, add an additional lens to bring the image the right way up and the telescope is at least 350 mm (14 in) long. Many old telescopes are much longer than that, in order to provide greater basic magnification. They are often 600 mm (24 in) or more in length. To make them more portable, they were often designed to collapse down using a series of draw tubes; or in other words, they telescoped. We still use the term even though few telescopes these days do this.

But imagine having two telescopes like this side by side—they would hardly be convenient. So the first binoculars, known as field glasses, used a different optical system, in fact the same type as used by Galileo in his first primitive telescope. Instead of having a magnifying lens for the eyepiece, they used a reducing lens, but inside the focal length of the objective. This has the effect of making the barrel shorter, and also gives an upright image, but with two serious drawbacks—they have a tiny field of view, and limited magnification. Many children's plastic binoculars are of this type, and you may still encounter them in theaters, where they are known as opera glasses. You put a coin into the holder to release a clip so you can use them for the duration of the performance,

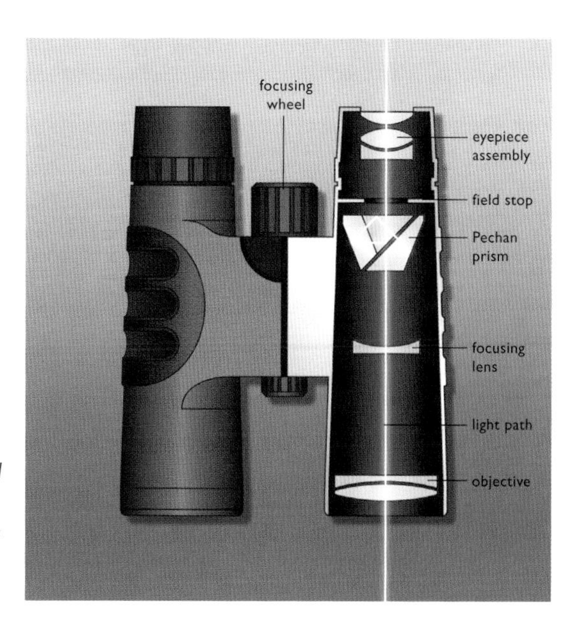

► *This roof prism instrument uses a Pechan-type roof prism to fold and invert the light beam. On this model, an internal lens is moved inside the barrels to provide focusing.*

and are rewarded with restricted views of the action on stage magnified about three times.

These are of limited use, so for better results enter, stage left, the true binoculars. The design was invented by an Italian, Ignazio Porro, in 1854. His clever idea was to use a pair of 90° prisms (known to this day as Porro prisms) to reflect the light inside the barrel back on itself then forward again, thus considerably reducing the barrel length. With a nifty orientation of the prisms he was also able to bring the image the right way up, thus avoiding the need for a further lens.

You may wonder why the image doesn't display massive amounts of false color—after all, prisms are noted for their ability to disperse light into its component colors. Actually, a 90° prism has a property known as total internal reflection—one surface acts as a mirror with no need for silvering. You can see the same effect underwater when looking up at the surface of the water at a shallow angle—the surface looks shiny, and you can't see objects above the water.

The more modern binoculars also fold the light, but using a more elaborate and expensive type of prism known as a roof or dach prism, dach being German for roof. This allows them to have what are apparently straight-through barrels, but in fact there are prisms inside each barrel. Pechan prisms are a similar alternative.

The prisms inside both types of binoculars, as well as being their strength, are also their weakness. You only have to dislodge the prisms slightly and the beams of light that emerge from each barrel will no longer be parallel or what is called collimated. The result—double images and what is usually referred to as double vision. The more expensive and robust instruments have well-secured prisms, that will withstand being put down rather forcefully onto the table, or even being dropped. But the prism fixings in cheaper models are quite likely to be easily knocked out of line. You may think that it would be an easy matter to open them up and recollimate them, but you would have to do this while looking through them, and work very precisely. Repairers have test rigs that allow them to do the job easily, but the cost of getting this done is usually greater than cheap binoculars are worth.

What are the advantages of each type? Porro prism binoculars are less expensive to construct, as the prism assemblies are cheaper, and the larger sizes of instrument are usually of this type. Roof prism binoculars need to be made to higher specifications, so are more costly, but they are considerably more compact, because the barrels are set less widely apart, and often weigh less. This makes them more popular in the smaller sizes of binoculars. There is no inherent drawback in roof prism binoculars for astronomy, except for a small additional light loss, and if they are a little lighter you may be able to use them for longer.

Focusing binoculars

The most common means of focusing binoculars is by center focus, with a wheel or sometimes a lever on the central pivot. In this case, both eyepieces are mounted on a separate bridge which moves in and out as you turn the wheel. This is convenient, but it is a source of problems in old or cheap binoculars if the bridge does not move smoothly in and out. If it catches or rocks slightly, you find that one eyepiece is constantly getting out of focus.

Some binoculars have independent focusing of each eyepiece, or focus by moving the front objectives. Independent focusing is not convenient for everyday use, particularly for such activities as birdwatching where your subject is frequently moving to different distances, but for astronomy where the subject is at infinity and only you are using the binoculars it is less likely to give rise to focusing problems.

Focus-free binoculars have the focus fixed, on the one-size-fits-all principle, which operates in the same way for binoculars as it does for clothes—fine, if they suit you. They are convenient for some purposes, such as for military and marine use, but they only suit people with standard eyesight—you can't even adjust the eyepieces for individual variations. They usually have fairly low magnifications, a typical specification being 8 × 42. Their great advantages are that there is less to go wrong, they are easier to waterproof than focusable binoculars, and they are instantly available for use, which is of benefit in a military environment, for example. If these appeal to you, it's essential that you look through them before buying, to make sure that they suit your particular eyesight.

Eye relief and exit pupil

These specifications are often overlooked by binoculars suppliers but can be fundamental to your ease of use, or even your ability to see through the binoculars in the first place. Eye relief is the distance between the rear of the eyepiece lens and the front surface of your eye. It depends on the design of the eyepiece, and in the case of the basic eyepieces used in standard 10 × 50s and 7 × 50s, the lower the magnification the longer the

▲ *These focus-free Konus 10 × 50 binoculars have no adjustments other than for interpupillary distance.*

◀ *The exit pupil of binoculars is the disc you see when holding them away from your eye.*

eye relief. Having long eye relief makes the instrument much easier to look through, as you don't need to press your eyes so closely to the eyepieces, and in particular allows you to wear spectacles. If you wear spectacles to correct for short or long sight you can usually focus the binoculars to suit your eyes without wearing spectacles. However, if your eyes have astigmatism, which makes a point of light appear as a short line, you may need to wear your spectacles when using the binoculars.

It is possible to design a high-power eyepiece to give improved eye relief, so if you do need to wear spectacles you might need to make this a priority. The other approach is to give the binocular objectives a longer focal length, so that the eyepieces need not be so powerful to give the same magnification. Many 20 × 50 binoculars, for example, have a particularly short eye relief and it can be difficult to see the full field of view. A better specification is 16 × 50 or 20 × 60. But this does not mean to say that magnifications higher than 20× or so are impossible to obtain—just that the instruments need to be designed with this in mind. So you can buy 30 × 60 binoculars which are fairly easy to use but some 30 × 80s which are not. Check the figure for eye relief before you buy.

Closely related to eye relief is the exit pupil of the binoculars. This is the diameter of the beam of light that emerges from the eyepiece, the bright circle that you see in the eyepieces if you hold the binoculars about 30 cm (12 in) in front of you. It varies in size with the size of the objectives and the magnification. To give some examples, in 10 × 50 binoculars the exit pupil is 50 mm (2 in) divided by 10, which is 5 mm (0.2 in). So you must place your eye exactly in line with this 5 mm

PUPIL SIZE VARIATION	
Age	Pupil size
19	7.0–8.0 mm (0.28–0.31 in)
40	6.0 mm (0.24 in)
65	4.8 mm (0.19 in)
74	3.0–3.4 mm (0.12–0.13 in)

(0.2 in) beam of light to see the full field of view. However, in 7 × 50s the exit pupil is just over 7 mm (0.28 in), and for 16 × 50s it is only 3.1 mm (0.12 in). This is a theoretical value, but as described later, in some poor designs you don't get the full exit pupil you are expecting.

Why is this important? The pupils of your own eyes vary in diameter with the brightness of the view. They are small—about 1.5 mm (0.06 in)—in bright sunlight, but they open up to a maximum within a matter of seconds in dim lighting. But this maximum size varies widely from person to person, and most strongly with age. The table above gives some typical values.

A young person can make full use of the 7 mm (0.28 in) exit pupil of 7 × 50 binoculars, and will get a bright image. But a 60-year-old given the same instrument will only be able to see 5 mm (0.2 in) of the exit pupil, reducing the effective aperture to only 35 mm (1.37 in). So the older person will get just as bright a view with a 7 × 35 instrument as a 7 × 50—and will find the lighter instrument easier to hold for long periods into the bargain. At the extreme, an elderly person with only a 3 mm (0.12 in) pupil will get no benefit from anything larger than 7 × 21 or 8 × 24 binoculars, and in bright sunlight anyone else will be in the same situation. In fact, 7 × 50 binoculars, which are often recommended for astronomy, are only really suitable for young people to use at night. But there are wide variations from person to person so it is worth trying to measure your pupil size if you can.

This is easier said than done, as experiments by author David Frydman suggest that the pupil does not open up fully until the light levels are very low. On this basis, if you get a friend to measure your pupil size with a ruler in subdued lighting, it will be approximately 10% smaller than its maximum.

You may wonder why, if your eyes open up to their maximum size in a few seconds in low lighting, your night sight improves over a period of 15 minutes or longer when you go out into the dark. This is because of chemicals that are produced within the eye that improve night vision, and is nothing to do with the pupil size. Incidentally, the use of drugs, including cigarettes and liquor, reduces your pupil size, so these are not recommended for astronomers.

Field of view

The size and magnification of binoculars are only the start of the range of variations possible. One of the most important is the field of view—and here you must distinguish between true and apparent field. The

FIELD OF VIEW		
True field	**m/1000 m**	**ft/1000 yd**
2.5°	44	131
3°	52	157
3.5°	61	183
4°	70	210
4.5°	79	236
5°	87	262
5.5°	96	289
6°	105	315
6.5°	114	342
7°	123	368
7.5°	132	395
8°	141	422
8.5°	149	448
9°	158	475

true field of view is the actual angle of view covered in degrees, such as 5.6°. In other words, if you regard the whole horizon as being 360°, you can see just 5.6° of it. It can also be stated as the width of view in meters at a distance of 1000 meters or the number of feet at 1000 yards. The number of degrees is of more interest to astronomers, as you often need to know how much of the sky is revealed by your own instrument. One tiny bit of sky looks very much like another. Refer to the table shown here to convert field of view in degrees into one of the other systems of measurement.

The apparent field of view is the impression you get when looking through the eyepiece—whether a narrow tunnel or a broad expanse. It is very nearly the true field in degrees multiplied by the magnification. In the example given above, if your 10 × 50 binoculars have a true field of 5.6°, the apparent field is about 56°. 50° fields are the smallest one should generally consider for astronomy—anything smaller seems like tunnel vision. True fields of 5.5° or 6° for 10 × 50 binoculars make viewing more enjoyable, and these are often described as wide-field binoculars. Nowadays 6.5° is considered extra wide. But during the 1970s, 7.5° was fairly common with some 10 × 50s having 7.8° or even 8° fields. Having extra-wide fields means that finding one's way around the sky is much easier, and chance events like passing artificial satellites or meteors are more likely to be seen.

With higher-power binoculars typical fields of view are about 4.5° for 11 × 80s, giving an apparent field of 50°, and 3.5° for 20 × 80s. A 3° field is probably the smallest that most people would find acceptable. The difference between ordinary and wide-field binoculars lies in the eyepieces, and not in the size of the objective lenses—as many people wrongly believe. Standard binoculars have three lenses or elements in their eyepieces, but wide-field eyepieces typically have four to six elements. This is an extra bulk of glass, which adds to the weight and of course the cost, but with modern lens coatings the light loss caused by the extra amount of glass is not really a problem.

At the low-power end you can get very wide true fields of view. The 4 × 21 Bushnell binoculars have 17° true fields—they are focus-free binoculars so you have to try them to see if they are suitable for your eyes. The 5 × 25 Russian Fotons are 12.5° but with a curved field—that is, not all of the field of view will be sharp at the same time and stars are not sharp at the edges. The 6 × 30 Russian extra-wide-angle binoculars have 12.5° fields. With such binoculars large areas of sky are visible at one time and they can be hand-held for long periods. This is useful for spotting artificial satellites, looking at comets, seeing faint meteors, or just general stargazing.

In most good binoculars the view is fairly sharp to near the edge of the field. In very few binoculars indeed is it really sharp to the edge. An acceptable standard is that the view should be sharp over at least two thirds of the field. In really poor designs only the central third of view is sharp. Such binoculars are not suitable for astronomy as their main purpose is to have a wide clear view, and binoculars with a limited wide field or a curved field can be very disconcerting to use as everything outside the central area looks like a comet or a nebula! Only when you bring the object to the center of the field do you realize that it is just another star.

One problem with wide-field eyepieces is that the angle of view may be reduced if you cannot bring your eye close enough to the eyepieces—particularly a problem when wearing spectacles. Eyepieces with long eye relief allow the whole field to be seen. Sometimes there are rubber eye-cups around the eyepieces which can also make it hard to get your eyes right up to the lenses.

"Movie" binoculars

When you look through binoculars, the view you see is virtually always circular. This is the natural shape of the field of view of any optical instrument, and if the instrument is working properly you should see only one circle. But almost invariably, movie makers do not believe that their audience will accept the view seen by someone using binoculars is a circle, and when showing a view as if seen through binoculars superimpose a mask with two overlapping circular fields of view. Every sales assistant who sells binoculars will have encountered a customer who asks why the binoculars they see don't give this field of view. Actually, it would be perfectly possible to make binoculars that did this, or had any other field of view including an apparent wide-screen letterbox shape, but the fact is that they would actually just be masking down the true circular field of view of the instrument.

Binoculars in which the optics were misaligned to such an extent that the two barrels would clearly be at an angle to each other, would indeed give a "movie" shaped field of view, but your eyes would not merge the two images into a single field of view.

Some binoculars have been made with novel fields of view, but they do not offer any real advantage over more conventional models. The reason is that the field of view that you see is defined by a mask inside the eyepiece, rather than having anything to do with the shape of the objective. You can test this by putting your finger in front of the objective while observing. All that will happen is that you will see a reduction in brightness over one part of the view. Your finger is completely out of focus. For the mask to be sharp, it must be at the focus point of one of the lenses in the eyepiece. So the mask is within the eyepiece itself. It is really only there to hide the internal construction and to give a neat edge. It could easily be replaced by any other shape of mask, though both eyepieces would need identical masks that stayed in an identical orientation, so special arrangements would be needed for adjusting the interpupillary distance.

How much detail can you see?

The ability of an instrument to show fine detail is known as its resolution or resolving power. Astronomers usually measure the resolution of a telescope by its ability to show two very close stars as separate objects. Without any optical aid, someone with good eyesight can separate two close stars if they are three arc minutes (180 arc seconds) apart, as long as the individual stars are not too bright or faint. So using good binoculars that magnify 10×, you would expect to be able to separate a double star just 18 arc seconds apart. This is a good average value, and applies where both stars are of a similar brightness. But sometimes the eye can do better than this because some binoculars have a smaller exit pupil than the average night-adapted eye. The popular Canon 10 × 30 image-stabilized binoculars, for example, have an exit pupil only 3 mm (0.12 in) across. Author David Frydman can just separate the double star Mizar at 14.4 arc seconds with these binoculars additionally supported but not tripod mounted. This is not an equal double—one star is five times brighter than the other. An equal double star would be seen at a closer separation, and acuity varies considerably from person to person.

In theory, unless they are of truly awful optical quality binoculars are capable of showing much finer detail than this, but only by using much higher magnifications. In practice, the amount of detail you can see is limited by the magnification of the instrument and the steadiness with which you can hold it. You can see both more detail and fainter stars if the binoculars are mounted in some way, as described on page 181.

The traditional resolution figures above apply to the closest separation of double stars. However, the eye can do much better than this when looking at surface detail on, say, the Moon. In this case, most people can easily see detail of one arc minute (60 arc seconds) for medium contrast detail and somewhat better for high contrast detail.

This is easily shown to be true when observing the Sun with fully protective filters (see page 193)—remember, do not view the Sun without proper protection as eye damage will occur. Great care is needed in solar observing. People with average vision can easily see sunspots 60 arc seconds across without optical magnification and skilled observers with excellent vision can see sunspots only 40 arc seconds across. This is even more apparent when viewing a rare transit of Venus—a very high-contrast event—with Venus being 60 arc seconds across. Many people can easily see the tiny black spot slowly crossing the Sun's surface using just a solar filter. And linear features can be seen on the Moon where the width is much less than these figures. Again, with 10× binoculars, you can do 10 times better than this, which equates to features only about 12 km (7 mi) across.

Zoom binoculars

There is nothing inherently wrong with zoom binoculars—even though many astronomers dismiss them. They can be very useful for giving high resolution on, say, the Moon or double stars, or for picking out fainter stars.

There can however be a number of problems with zoom binoculars. They tend to be collimated at the factory at the lowest magnification instead of at the highest, with the result that few of the cheap or modestly priced ones are ever properly collimated, which in turn means that at the higher magnifications you get unavoidable double vision. Some models, usually sold by mail order, have too large a zoom range, with in some cases an insanely high top magnification, such as 100. A practical limit is about 1 per mm (0.04 in) of aperture—so 25 for 25 mm binoculars is acceptable, but 100 for 30 mm binoculars is not. At such high powers the image becomes impossibly dim, and the slightest movement has magnified impact.

In order to have a zoom function extra elements are required within the eyepieces, so the field of view is smaller and there may be more distortion in the image than with a fixed magnification. However, with modern lens coatings the extra glass should not absorb a significant amount of light, which was one criticism of zoom eyepieces in the past.

While cheap fixed-power binoculars can often work well, in the case of zoom binoculars there are more moving parts to go wrong and a greater chance of something not working properly. Bear in mind that the two eyepieces have to work together and give equal magnifications, and if the zoom lever is flimsy or jerky you may find that the zoom function is hard to use successfully.

Modest range zooms (up to 3×) of good construction can perform very well. Be wary of binoculars with a zoom range above three and a maximum magnification above 30 times in small binoculars (say 10–30 × 30) or 40 times in larger sizes (say 12–36 × 70). Good examples are the Nikon Action 10–22 × 50 or the Nikon 8–24 × 25 Travelite.

The contention that the extra lenses required by zoom eyepieces will give a dimmer view is not in itself significant with modern coatings. Some only have one extra lens element per barrel, and this is used to good effect for better close focusing and better edge performance in addition to the zoom function. It has to be admitted, though, that there are many more poor zoom binoculars available than good ones. You should test them very carefully at all magnifications before purchase.

Weight

The weight of a pair binoculars is one of the most important factors affecting the length of time they can be hand-held. In general the weight is decided by the size of the objectives. Here are some average weights of different aperture binoculars:

WEIGHTS OF BINOCULARS			
Mag.	**Aperture**	**Weight**	**Comments**
6–10	30–32 mm	500–650 g (17.5–23 oz)	Can be hand-held for long periods
8–10	40–42 mm	600–850 g (21–30 oz)	
7–12	50 mm	750–1000 g (26.5–35 oz)	
12–20	60 mm	1100–1400 g (39–49 oz)	Can be hand-held for short periods
10–20	70 mm	1400–2000 g (49–70 oz)	Short periods for the lighter ones. Tripod needed otherwise
11–20	80 mm	1950–2500 g (69–88 oz)	A tripod usually needed

Some high-quality binoculars, particularly in the 50 mm size, weigh more than shown here. A 50 mm pair can weigh 1200 g (42 oz) or in the case of some 7 × 50 rugged binoculars with internal compass 1650 g (58 oz). Everybody varies widely in their ability to comfortably hand-hold a pair of binoculars without tiring, and each person needs to find out their limits. The degree of steadiness also varies from person to person, and it may be this rather than comfort which is the deciding factor.

Lightweight binoculars may not necessarily be a good thing. Cheap models save on weight by using a lot of plastic in their construction, and are likely to have a short lifetime in active use. But binoculars intended for nature viewing can be designed to be light in construction without compromising on quality, and if this is important to you, you may have to pay more for it. Such binoculars will probably also be waterproof and have a more rugged construction.

Image-stabilized binoculars

The big drawback with binoculars has always been the difficulty of holding them steady. When you are trying to see fine detail, the slightest movement of your arms becomes magnified several times, making the job much harder. Technology has now come to the rescue with the introduction of image-stabilized (IS) binoculars, which use either an electronic or a mechanical system to keep the image steady.

The difference can be striking in use. Gaze at your chosen star and it is all over the place as your arms move. Press a button and the movement all but stops. The system does allow you to move the binoculars from one spot to another without trying to lock on to objects, but it is most sensitive to the sort of natural movements that you want to counteract. This calls for a careful choice of what is called the compensation angle—the amount of shift in viewing angle that the instrument will correct for. The Canon choice of 1° appears to be a good compromise.

There are several systems for image stabilization. Canon's uses electronic sensors that detect movement and actively shift an optical component in the light path, and it therefore requires two AA batteries for operation. You press a button and the image stabilizes almost immediately, overcoming hand-shake very successfully though less so for overall arm movement. You may need to rest your arms on a solid object for a really motionless view.

The Fujinon system includes a gyro attached to a prism, which therefore maintains its attitude despite being moved. Batteries are also required for this system, and there is a two-stage operation. The first press of the button spins up the gyro, which takes a second or two to reach full speed, and a second press then stabilizes the binoculars. Reports suggest that it overcomes arm movement better than the Canon, but at the cost of difficulty in panning across the view. Some motor vibration has been seen on the stabilized image.

A third system, used on some Russian-made stabilized binoculars, requires no batteries at all, but instead uses a floating prism on a gimbal. Pressing a button simply unlocks the prism, and some say that it works well, while others are not convinced. This is the heaviest but cheapest system, and has the added drawback that in order to provide a steady image only part of the full aperture of the binoculars is used at any one time, so the nominal 50 mm aperture is reduced to only 35 mm.

So how useful are image-stabilized binoculars? They are not cheap, but in particular the Canon 10 × 30 binoculars, which we have used, offer exceptionally good image quality, with almost pin-sharp images from edge to edge of a very wide field of view, as a result of a field-flattening component. Unlike non-stabilized binoculars, all IS

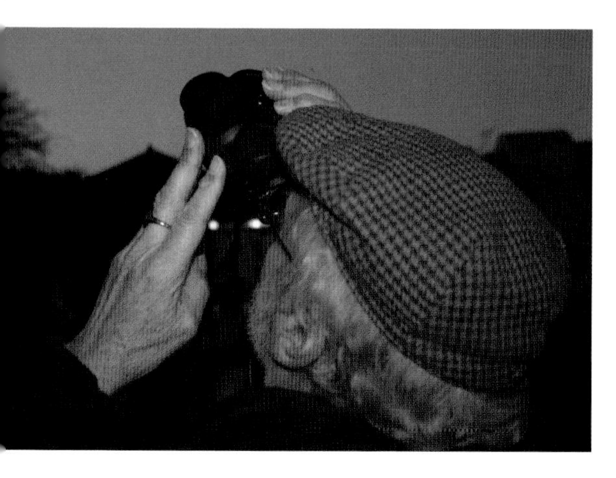

◀ *Canon 10 × 30 image-stabilized binoculars are compact yet give superb performance.*

binoculars offer higher magnifications than normal for the aperture. They may also be heavier than other binoculars of the same aperture, but there is a wide variation in weight in binoculars anyway depending on the degree of inbuilt ruggedness.

The steadiness of the image means that you can detect faint objects and see fine detail much more readily with IS binoculars than with conventional hand-held models. Author David Frydman claims that his Canon 10 × 30 binoculars outperform any standard 10 × 50 hand-held models, and gain a magnitude in the faintest star visible. They have convenience of use, as you are not constantly trying to find some nearby object against which to steady yourself. If the IS system were not expensive, it would probably be the popular choice.

Lens coatings

It may seem strange that by actually coating a layer of some material onto a piece of glass you can increase the amount of light that it transmits, but this is exactly what coatings do, and they can make a dramatic difference to the brightness of the image.

Each time light crosses from air into glass or vice versa, a little light is reflected instead of passing straight though—around 4% or 5%. There are more than a dozen such transitions in even the simplest of binoculars, and as a result uncoated binoculars may only transmit about 50% of the light and also suffer from a loss of contrast. Where two lenses adjoin, such as in eyepieces and objectives, it has always been common practice to cement the two lenses together, as this improves the light transmission. Coating was introduced to binocular lens elements in World War II, though the early coatings were sometimes soft and easily rubbed off, particularly on the outer surfaces of the eyepieces. Slowly, many-layered or multicoating was applied and by the 1980s had reached very high standards in top-quality binoculars.

The best binoculars today can have a total light transmission of 95% or more. Porro prism binoculars have a slightly higher transmission than the equivalent roof prisms. With great care in baffling (internal black rings) and blackening the edges of lens elements—both to stop stray light—the contrast can be very high.

The problem with coatings lies in the definition of the term. Binoculars are designated coated, fully coated, multicoated, or fully multicoated. There is no standard and among cheaper binoculars a lot of license is taken in these descriptions. Fully coated or multicoated can mean that only the two exterior surfaces of each barrel are single coated. With the best brands no statements are made at all, yet every air/glass surface including the prisms is carefully given multilayered coatings. In some cases even the cemented surfaces are additionally coated.

With really cheap binoculars the transmission is only 60% and may be less if the glass is slightly gray instead of clear. In addition, internal vignetting (see page 165) can cause 10 × 50 binoculars to have exit pupils of only 3.9 mm (0.15 in) instead of 5 mm (0.2 in). The total light loss can be very high. In the worst cases only 30% of

▲ *Two different lens coatings. Left, ruby coating on Konus Sportly 10 × 50s; right, Pentax XCF 10 × 50s. The Konus binoculars have a bluer tinge to the view as some red light is reflected by the strongly colored coating.*

the light is transmitted even when new. As a result, very good 40 mm binoculars can transmit more total light than a very poor 60 mm example. On the other hand, some very reasonably priced binoculars such as the 15 ×70 Revelation or the 12 × 50W Minolta Activa have very high transmission.

A further slightly ridiculous cosmetic coating is the so called ruby coating which is applied to the outer front lens element. It is claimed to give increased visibility in hazy conditions. In author David Frydman's experience it shows a blue-tinted world with high light loss. It is to be avoided for astronomy. For birdwatchers it seems to have the undesired effect of frightening the birds as they see two large, red eyes staring at them and, of course, they fly away.

You can check for yourself the extent to which a particular instrument has coated optics. Look into the objectives with a light behind you and you will see several reflections of the light, mostly diminished in size compared with what you would expect from a flat surface. If these are colored to any extent, the surfaces have been coated. The color can vary from light straw to strong green or red. The prisms will reflect light as flat surfaces, so you may be able to tell whether they have also been coated. Also look into the eyepieces in the same way.

Every manufacturer has its own name for their coating method, and you will often see this trumpeted in their publicity.

Prisms

The glass from which binocular prisms are made can have an effect on the brightness of their images. The glass used for optical components is considerably different from ordinary window glass, and different

A range of models

The selection of binoculars shown on these pages is only a guide to the range available, and shows the variations in each price range. It is not a list of recommended models, though all of them should be useful for astronomy. The specifications are taken from manufacturers' data, and have not been measured independently.

Prices vary widely with the supplier and from country to country, so we have compared them with typical prices of other consumer goods. The price codes are on a scale where A = low and X = very high. As a rough guide, A would be the cost of a cheap zoom compact film camera; B a cheap digital camera; C a low-cost 15" television or a reasonable digital camera; D a higher-quality 15" to 19" television or high-spec compact digital camera; E = a 20"–23" television or mini-DV camcorder; F = a 23"–26" television or consumer digital SLR; X = a large-screen television.

additives are used to control the amount by which a particular glass refracts light. Optical glass is manufactured at only a few specialized factories, so it is quite likely that binoculars from the Far East will use German Schott glass in their lenses and prisms, though some of the major camera manufacturers, such as Nikon, do make their own glass. A widely used material is BK7 glass, which is a borosilicate crown glass. A more expensive glass is BaK4, which tends to be used in higher specification models. Because of the higher refractive index of BaK4, the exit pupil appears circular except possibly in very wide-angle instruments, whereas with BK7 glass the prisms need to be much larger to achieve this.

Faults with binoculars

Whether you are choosing or testing binoculars, there are many points you have to bear in mind. These fall into the general categories of optical defects and mechanical defects, of which the second can affect the first.

Optical defects

Your first look through binoculars can be very impressive. Distant objects appear close, you can spot details that you never realized were there, and the scene appears bright and sharp. But take a closer look and you may start to see defects that will have a considerable effect on your viewing pleasure.

False color or chromatic aberration is a common problem. This shows up as a colored fringe around sharp edges of objects. Virtually all binoculars have some degree of chromatic aberration, but it should not be noticeable in a general view and should only appear when you are

Canon 10 × 30 IS Porro

Weight	610 g (21.5 oz)
True field	6°
	105 m/1000 m
	315 ft/1000 yd
Exit pupil	3 mm (0.12 in)
Price code	E

Comments

Brilliant for astronomy despite small aperture. Crisp star images across whole field. Also 12 × 36, 690g (24 oz), 5° field.

looking at an object against a bright background, usually bright sky. If it shows up badly by day, it will appear worse by night when you are looking at the Moon. It is sometimes said that the most acceptable colors for chromatic aberration are apple green, and plum, while vivid blue and orange are least desirable.

Coma is seen toward the edges of the field of view, and is most noticeable on star images or, failing that, on bright points of light such as reflections off railings. It causes bright points to appear V shaped like seagulls, with haze between the arms of the V. Again, if it is obvious by day, it can turn a starfield into a whole host of comets. Poor edge performance is unfortunately a fact of life for astronomers. If stars are fairly sharp to two thirds of the way to the edge that is acceptable. Four fifths is much better. If your binoculars do better than that you are doing very well. But really there is no justification for the general poor edge performance in binoculars. With modern sophisticated design the optics could be much improved as Canon have shown with their image-stabilized binoculars.

Astigmatism also appears on star or point images, though it may make the overall view seem somehow unsharp. It is caused by a lens that is not perfectly symmetrical, so that its magnification is greater along one axis than another. A star image therefore appears as a short line, which changes orientation as you try to focus it. But beware—your own eyes may be slightly astigmatic, so if you see this fault carry out a simple test. Look through just one half of the binoculars and rotate it as you view. If the astigmatism rotates with the optics, then the instrument is at fault. But if it remains constant no matter what the orientation of the instrument, your eyes are the problem.

Olympus 8 × 40 DPSI Porro

Weight	710 g (25 oz)
True field	8.2° 143 m/ 1000m 430 ft/ 1000 yd
Exit pupil	5 mm (0.2 in)
Price code	A

Comments

Good-value general purpose basic binoculars from a well-known camera manufacturer.

Curved fields occur when different parts of the field of view have different focus points. This is surprisingly common, and is obvious if you have to refocus an object when it is at the edge of the field compared with the center. Though it is undesirable for astronomy, if can actually be an advantage for nature study if objects in the foreground are in focus at the same time as more distant objects at the center of the field of view, though whether this in intended by the manufacturer is another matter.

Field distortions are obvious when you are looking at a straight line such as a roof, which can appear curved. Pincushion or barrel distortion, whose names refer to the shape that a square would take on, are commonly seen. These distortions are not particularly troublesome in astronomy, but in extreme cases they can be irritating for terrestrial use. Another type of distortion is a result of varying magnification across the field of view. This is noticeable if you pan across a regular surface such as a brick wall, as the bricks seem to grow and shrink depending on where they are in the field of view. Again, it is more disturbing in terrestrial use than in astronomical observation.

In old binoculars, it is common to find that the cement between optical components has started to decay, resulting in a misty appearance. There is nothing to be done about this unless the instrument is so valuable that it is worth having the fault corrected by an experienced optical worker.

Quoted specifications are sometimes in error, even in good quality instruments. Author David Frydman has come across good 20 × 70 binoculars where the true field was 25% less than marked. This seems to be have been an error rather than a deliberate attempt to mislead as the binoculars are actually very good. Magnifications can

Bushnell 8 × 42 H2O Porro

Weight	770 g (27 g)
True field	8.2° 143 m/ 1000m 430 ft/ 1000 yd
Exit pupil	5.25 mm (0.21 in)
Price code	A

Comments

An economically priced waterproofed, and multicoated general purpose instrument.

be up to 10% in error either way. A simple way to test the magnification of an instrument is to look through one half of it only at a regularly repeating pattern, such as a railing or brick wall. You have to superimpose the view through the optical system on top of your direct view, and count the number of railings, say, in the direct view compared with the magnified view.

In *Chapter 6* we give the separations of some easily located stars in degrees. You can use these to check the actual true field of view of your binoculars.

Mechanical defects

In this category is included not just the mechanisms of focusing and so on, but just as importantly the way the optics are mounted.

Collimation refers to the alignment of the two optical paths, which should of course be parallel for the two images to appear together as one when you look through the binoculars. Poor collimation, giving rise to double vision, is where many binoculars, both new and used, come to grief and is the most likely reason why binoculars are not fit for use. The whole point of binoculars is to have two sets of optics that give merged restful images. The mechanical axes of the two barrels and the hinge line should be parallel. In addition, the two optical axes should be parallel to the barrels. You may be able to put up with a small amount of miscollimation at low magnification, but at high power even a tiny amount can be disastrous.

In cheaper binoculars the tolerances and assembly are often poor with the two barrels and hinge line not exactly parallel. When the optics are collimated at the factory, the only way to get the two optical trains

Minox 8 × 42 BL BR Roof

Weight	605 g (23 oz)
True field	6.5° 113.4 m/ 1000 m 342 ft/ 1000 yd
Exit pupil	5.25 mm (0.21 in)
Price code	E

Comments

A lightweight product from a manufacturer specializing in miniature cameras. Watertight to 5 m (16.5 ft).

in line is by offsetting the optics considerably to compensate for the mechanical imbalance. In severe cases this results in a cutoff of one of the fields of view, making the binoculars unpleasant to use. The circular outlines of the fields of view may not coincide even though a central object can be viewed correctly merged.

Another problem is that collimation may only be correct at one interpupillary distance. This may only become obvious when someone else, such as a child, uses the binoculars, and has to adjust the separation. If collimation is correct at a setting of, say, 64 mm (2.5 in), when the two halves of the binoculars are rotated to say 60 mm (2.3 in) correct collimation is lost.

A major source of poor collimation in binoculars is that the prisms get dislodged by a knock—maybe, if you are unlucky, before they even reach you if you are getting them by mail order. Some of the cheaper binoculars—even though cosmetically beautiful with rubber armor and contoured finger holds—are considered by some as "one use" binoculars. This is a bit unfair—but any serious knock may leave them useless and not economically repairable unless under warranty.

With medium quality binoculars, experience shows that about 10% are imperfectly collimated when sold, but some of these may be usable if the collimation error is small and in the horizontal plane. The observer's eyes may compensate as they do in reading when they naturally angle together. Collimation errors are less tolerable if they are at right angles to the eye line.

In top quality binoculars, collimation errors are rare as great care is taken to avoid such problems. They can stay in collimation for years of hard use.

Swarovski 10 × 42 El Roof

Weight	780 g (27.5 oz)
True field	6.3° 110 m/ 1000 m 330 ft/ 1000 yd
Exit pupil	4.2 mm (0.17 in)
Price code	X

Comments

Austrian manufacturer of high repute. Rugged and suitable for a lifetime of use in the field. Close focus of 2.4 m (7.8 ft).

◀ *Some cheap binoculars show squared-off exit pupils, which means that the prisms are restricting the effective aperture of the binoculars.*

Other mechanical faults include rocking or sticking bridges. In the case of a rocking bridge, as you focus the relative distance between each eyepiece and its objective does not change evenly, so the two halves of the instrument are focused slightly differently from each other. The result is that you find yourself constantly twiddling the diopter focus eyepiece to get a good view. Quite possibly the very act of pressing the binoculars to your eyes is enough to make the relative focus positions change, which leads to constant refocusing.

Some instruments have stiff focusing mechanisms, which is not a major problem if you use them only yourself and for astronomy, but is irritating for terrestrial use.

Swift 8.5 × 44 Audubon Roof

Weight	664 g (23.5 oz)
True field	6.4°
	112 m/1000 m
	336 ft/1000 yd
Exit pupil	5.2 mm (0.2 in)
Price code	E

Comments

This specification is regarded as ideal for birding. A roof prism version of a Porro prism model with a very high reputation.

Lack of ruggedness in manufacture is a common problem with cheap binoculars. This may not matter as much for astronomy as it would for nature study or marine use, where the instrument is far more likely to be subjected to an adverse environment or rough use, but it can only take a slight jar to knock the whole system out of collimation. It is easy to simulate the appearance of ruggedness, with molded exteriors and khaki-colored plastic, the real question is whether the optical components are firmly fixed, all the joints are waterproofed, and the outer glass surfaces given an overcoating which will withstand repeated cleaning, as frequently happens with marine binoculars which quickly and unavoidably get coated with salt spray. Even if you just live by the sea this can be a problem.

The mounting of the internal components such as the prisms can at times obstruct the light path through the instrument to a certain extent. If you look into the objectives, you may see squared-off edges within the instrument. The effect of these is the reduction of the effective aperture of the objectives—what is called vignetting. You will not notice this in use, but it means that you are getting less light through than you should.

Some 50 mm binoculars may in reality be as small as 39 mm, giving a light loss of almost 40%. If, in addition, they have coatings on only the two exterior surfaces—transmitting only 60% of the remaining light—the total light transmission is 36.5%. Furthermore, if poor quality slightly gray glass is used for the lenses or prisms the resulting overall transmission is maybe 30%—a 70% light loss!

Total transmissions of less than 50% are fairly common in cheap binoculars. Compare this to top quality fully multicoated Porro prism

Baigish 12 × 45 Porro

Weight	850 g (30 oz)
True field	5.3°
	92 m/1000m
	277 ft/1000 yd
Exit pupil	3.75 mm (0.15 in)
Price code	A

Comments

Russian binoculars of low price. Rugged model, though of basic design. Good for use in light polluted areas where the higher magnification helps to show deep sky objects.

binoculars that transmits 95% of the incoming light. The corresponding figure for top quality roof prism binoculars is 92%.

Making your choice

You will now understand that actually looking through binoculars before you buy them is a great advantage. But before you even start testing models, it is a good idea to decide what sort of instrument you want. The choice is enormous. There are over two thousand different models of binoculars available. Many of these are badge-engineered—that is, the same binoculars are sold under several different names. Even so, there are still over one thousand genuinely different models available, which makes a choice difficult for the consumer.

To start with, narrow down your choice to the sort of size you are interested in. Do you want a general purpose instrument that you can take everywhere, carry on walks in the countryside and to the beach, and which will show you a few objects in the sky as well? Or are you after something that will primarily be good for astronomy, but will be useful to carry in the automobile for occasional use? Or have you in mind something that will really bring in those deep sky objects, and which will mostly be used for serious observing? Perhaps you have another hobby for which the binoculars will also be useful? So to begin with, here is a description of the various sizes and types of binoculars, with comments on their usefulness for astronomy.

Small binoculars

In this category fit a wide range of general purpose binoculars, ranging in aperture from 24 mm up to 42 mm, with magnifications rang-

Zeiss 12 × 45 DPS Conquest T* Pechan

Weight 605 g (23 oz)
True field 4.6°
80 m/1000 m
240 ft/1000 yd
Exit pupil 3.75 mm (0.15 in)
Price code F
Comments
Watertight, nitrogen filled. T* (T-star) refers to the multicoating method. Less-expensive binoculars from a top quality manufacturer. Also 10 × 45 and 15 × 45.

▲ *Amateur astronomer Dave Tyler can hand-hold Revelation 15 × 70 binoculars. He has folded down the rubber eyecaps so that he can wear his spectacles while using them.*

▲ *His son Tom Tyler is happier with Nikon 10 × 35s which are half the weight of the 15 × 70s and have a field of view of over 9° compared with 4.4°.*

ing from 6 to 10. For the astronomer, they will provide good views of the Milky Way and many star clusters, and give excellent views of the Moon. The lower-power versions will not reveal the smaller and fainter deep sky objects very well in average conditions, however, so they are not regarded as the ideal binoculars for all astronomical purposes. Nevertheless, their lighter weight and convenience of use puts them firmly into the category of "the best instrument is the one you use most."

Pentax 10 × 50 XCF Porro

Weight	860 g (31 oz)
True field	6.5°
	114 m/1000 m
	340 ft/1000 yd
Exit pupil	5.0 mm (0.2 in)
Price code	A

Comments

Durable and reasonably priced binoculars from a famous camera manufacturer. The classic astronomy specification. Also 12 × 50 and 16 × 50.

They are often used by birdwatchers as they are comparatively light, yet give bright views by day. Binoculars for birdwatchers and for nature study tend to be waterproof and nitrogen filled, which is less necessary for astronomers. As a result birdwatchers' binoculars are often more expensive, typically costing around the same as a small good-quality television, whereas an astronomer would expect to pay about half that. Some birdwatchers opt for luxury non-stabilized binoculars costing as much as the larger widescreen televisions, and these are rarely found among the astronomical fraternity.

Birdwatchers generally use somewhat lower magnification than astronomers. So a typical birdwatching instrument is a pair of 8 × 42s, ideally with more waterproofing than an astronomer would need, but smaller 8 × 32s might be alternatives for their lighter weight. Of the two, 8 × 42s are more suited to astronomy, particularly in dark skies, and 10 × 42s are better still. In the image-stabilized range are Canon 8 × 25s, which are rather small for astronomical use but which might be suitable for elderly people who have small pupils, prefer light binoculars, and may suffer from shaking hands. The very popular Canon 10 × 30s and the more powerful 12 × 36s at twice the price are also widely used. Gary Seronik, whose "Binocular Highlights" column in *Sky and Telescope* magazine has been running for many years, says that his Canon 10 × 42 image-stabilized binoculars are his favorite instrument. But their price is about the same as very large widescreen televisions.

Medium-sized binoculars

The 45 mm and 50 mm sizes can be ideal for many astronomical purposes. They come in a wide range of magnifications, commonly 7 × 50,

Celestron OptiView 10 × 50 LPR Porro

Weight	850 g (30 oz)
True field	7.0°
	122 m/1000 m
	367 ft/1000 yd
Exit pupil	5.0 mm (0.2 in)
Price code	B

Comments

Drop-down light pollution filters help reduce sky glow, but don't expect a dramatic difference.

8 × 50, 10 × 50, 12 × 50, 16 × 50, and 12 × 45 for hand-held binoculars, with Canon 15 × 50 and 18 × 50 image-stabilized binoculars.

The 7 × 50 and 8 × 50 models, while often recommended as night binoculars, can only be used to full advantage by people whose pupils will open up to 6 or 7 mm (0.24–0.28 in), which usually means those under about 40. While they give the brightest views of the sky, this is not always ideal. Many objects, such as galaxies and the smaller star clusters, require magnification as much as aperture. There is a school of thought that claims that it is the magnification multiplied by the aperture that is the best indicator of binoculars' usefulness in astronomy. So 7 × 50s, giving a product of 350, would be beaten by 12 × 45s, giving 540, a finding confirmed by both authors. The latter will even beat 10 × 50s in many cases, on account of their higher magnification. Author Robin Scagell finds that the galaxy M82 is more readily visible in inexpensive but good Russian 12 × 45 binoculars than in more costly Pentax 10 × 50s, for example. And although the higher magnification of 16 × 50s gives a dimmer view than 12 × 50s, the increased magnification can actually make it possible to view some faint objects. Bear in mind, however, that binoculars with magnifications greater than 10× become increasingly difficult to hand-hold without a support or image stabilization.

As well as their use for astronomy, binoculars of this size are widely used for marine and military purposes. The most used yachting and boating binoculars are 7 × 50 waterproofs with or without an integral compass. They sometimes have a simple rangefinder, and the compass can be lit. Often a floating strap is used or the binoculars themselves float. Yellow is the traditional high-visibility color but nowadays black,

Steiner 7 × 50 Skipper Porro

Weight	1055 g (37 oz)
True field	6.7°
	118 m/1000 m
	354 ft/1000 yd
Exit pupil	7.1 mm (0.28 in)
Price code	D

Comments

Waterproof, nitrogen-filled. Steiner binoculars are widely used by the police and military. Model with illuminated compass also available. 48° apparent field.

green, blue, and other colors are made. Some are submersible in a range from 1 m (3.3 ft) to Steiner's 10 m (33 ft). The professional versions are fairly heavy from 1000 to 1650 g (35–38 oz) but lighter, less robust ones are also available. Yachting binoculars usually have independent eyepiece focusing.

One reason why 7 × 50 binoculars are favored for marine and military use is that because the exit pupils are much larger than the average daytime pupil size, the positioning of your eyes against the eyepieces is not so critical as with smaller instruments. If the boat rocks or the helicopter vibrates and your eye shifts slightly, you will still be able to see through them. Some military binoculars, which you might find secondhand, include a graticule or scale in the field of view. Graticule binoculars are not a good idea for astronomy, as the scale, as well as any dust that falls on the glass that it is engraved upon, can get in the way of stars.

The Fujinon image-stabilized binoculars are designed for marine and heavy professional use including helicopter observation. They are fairly heavy in the 14 × 40 version, so are included in this category. Canon stabilized binoculars are less used as most models are not waterproof, although the latest 10 × 42s are waterproof, and the 15 × 50s and 18 × 50s are described as all-weather.

The local racetrack is also the place to see medium sized binoculars. It is said that high-powered image-stabilized luxury binoculars were originally designed for observing the far side of horseracing circuits. If so, astronomers have to be thankful, as image-stabilized binoculars are a real step up in performance compared with even the best standard binoculars.

Helios 10 × 50 Ultimate HR Porro

Weight	790 g (28 oz)
True field	5.3°
	92 m/1000 m
	277 ft/1000 yd
Exit pupil	5.0 mm (0.2 in)
Price code	C

Comments

Japanese-made. High optical performance at a modest price with a good specification for astronomy.

Large binoculars

Here we come into the seriously heavy instruments, which you would not consider hand-holding for any length of time. They include 20 × 60, 15 × 70, 20 × 80, and 25 × 100 binoculars; considerably larger sizes are also available. They need not cost a great deal. Russian 20 × 60s are available for around the cost of the cheapest digital cameras, and Chinese-made Revelation 15 × 70s cost even less. These are not rugged binoculars which will stand up to a lifetime of abuse before the mast, but with care should perform well for astronomy.

Author Robin Scagell has found in comparative tests that 20 × 60s will easily reveal difficult objects such as the Crab Nebula that are completely invisible in 10 × 50s or 12 × 45s. For locating faint deep sky objects, studying the Moon, and most astronomical purposes, such instruments excel. It is worth noting that at this level a telescope of similar or larger aperture will give you more flexibility regarding magnification and should be considered for its potentially lower cost and no greater encumbrance. However, one thing a telescope will not do is to give you that authentic binocular view, using both eyes to search out your prey, unless you use a specially designed binocular viewer.

Such instruments are often used for coastal observation, and they grace the bay windows of many seaside homes. If you do require them for all-weather use, however, you can expect to pay accordingly. Quite often, these instruments have angled eyepieces for ease of viewing, but bear in mind that finding heavenly bodies can be quite a task in this case as you are not viewing in the same direction as the object. Angled-eyepiece binoculars are often provided with a sight for astronomical viewing which is only of real use if they are tripod-mounted.

Bresser 10 × 50 Corvette Porro

Weight	830 g (29 oz)
True field	5.1°
	89 m/1000 m
	267 ft/1000 yd
Exit pupil	5.0 mm (0.2 in)
Price code	A

Comments

One of the better models from the cheaper end of the market.

Whatever the size of binoculars you buy, you may well wish at some stage to use them on a tripod. Details of the way binoculars are mounted on tripods are given in *Chapter Six*, but bear in mind the method of tripod mounting when choosing the right model for you. This is rarely mentioned in specifications, and yet it could play a major part in the ease of use of the instrument.

Sources of binoculars

Having taken into account all the above considerations, namely the weight of the binoculars, your interests, and even your pupil size, you are still faced with thousands of different models to choose from. There are instruments from the major camera manufacturers, such as Nikon, Pentax, and Canon. There are other optics manufacturers, such as Swift and Vixen. There are "own brand" models that you encounter only in a particular chain of retailers. Many of these are badge-engineered—that is, the same binoculars are sold under several different names. So which should you aim for?

In general, you can't go too far wrong with a known brand, but even some of the best manufacturers have produced quite costly binoculars with faults such as seriously curved fields, something most people would find it hard to live with. At the same time, there are some discounted and undoubtedly cheap binoculars, usually made in China despite the European-sounding name on the label, that give superlative performance for astronomy. But bear in mind that two instruments with apparently identical specifications may be quite different in performance. Optical components are not stamped out identically by a press—they require some hand-finishing, and each plant can turn out

Leica 12 × 50 Ultravid BL Roof

Weight	1040 g (36.5 oz)
True field	5.7°
	100 m/1000 m
	299 ft/1000 yd
Exit pupil	4.2 mm (0.17 in)
Price code	X

Comments Magnesium body, titanium hinge, ultra-rugged, waterproof to 5 m (16 ft). −25°C to +55°C (−13 to 131°F). Bright, high-contrast images, fine color fidelity

good and bad lenses or prisms that look identical until tested. It is quite feasible to separate out the low-grade components from the high-grade ones, and to produce binoculars in identical-looking units that are actually widely different in performance.

The vast majority of the millions of binoculars sold each year are probably Chinese-made. One Chinese factory alone claims the capacity for producing 800,000 units a year. The quality of Chinese binoculars as a whole covers the full range from top to bottom. During the Soviet era, Russian binoculars were widely available, providing basic and rugged instruments at a low price, but these are now less widely available. However, they are still imported and are found under a variety of trade names.

Good brand-named binoculars, for example from the large camera makers, are offered at discounts fairly frequently. Sometimes these are slightly older models, which are little different from current models. Sometimes they are with specifications that don't sell quickly. Often these specifications are particularly suitable for astronomy, but not ideal for general daytime use. In many cases they are Porro prism binoculars, which are cheaper than equivalent roof prism models. The public prefer the roof prism types—and leave astronomers the ideal Porro prisms at a discount on an already lower price.

Where and how to buy

You have a wide range of options these days as to where you actually buy your binoculars. The only source that we would not recommend, or at least would be very wary of, is the mail-order supplier advertising special purchases of binoculars in newspaper advertisements or similar,

Nikon 12 × 50 CF Action EX Porro

Weight	1045 g (37 oz)
True field	5.5°
	96 m/1000 m
	288 ft/1000 yd
Exit pupil	4.2 mm (0.17 in)
Price code	D

Comments

Medium priced binoculars from leading camera manufacturer. 8 × 40 and 10 × 50 models also available.

who is not apparently a specialist supplier of quality binoculars. In this case the motto is "Buyer Beware!"

By far the best method is to go to a supplier with a wide range of different binoculars in stock so that you can compare different models side by side. Inevitably, you will pay a bit more for the facility than if you were to order from an Internet-only supplier or by mail order. It's for you to weigh up the value you place on the facility and also on such factors as the ease of getting a replacement or repair should you find a fault.

One drawback with testing binoculars at point of sale is that if you are intending to use them for astronomy, a daytime test is not really suitable. You can look at all the rooftops and railings you want, but during the daytime your pupils will probably be closed down to 3 mm (0.12 in) or less on a bright day, so you will only be using the central part of the optics. The view on a dark night might be quite different. Ideally, you should test on a dull day, or on a winter evening when it gets dark early. If you can, visit on a clear evening and when there is a planet visible (though bear in mind that Saturn and Venus are not usually circular). The Moon is a good alternative, which you may be able to see if necessary in daylight.

Even if you have to choose by day, a practical test will tell you much more about the instrument than any specifications, so it is well worth making the effort to test before you buy. But before you start to make comparisons, familiarize yourself with the correct way of looking through binoculars (see page 178). In a nutshell, get the interpupillary distance correct, focus using the left half, then focus on the same object using the diopter focuser. The first of these stages is particularly impor-

Celestron 15 × 70 Skymaster Porro

Weight	1361 g (48 oz)
True field	4.4°
	77 m/1000 m
	230 ft/1000 yd
Exit pupil	4.7 mm (0.18 in)
Price code	B

Comments

Similar instruments available under other names, such as Revelation.

tant. If you don't have the binoculars set for the distance between your eyes, you will not be looking through the center of each optical train. Errors in optical systems increase as you move away from the centerline, or optical axis, of the system. Bear in mind also that if you have minor defects in your eyesight, such as astigmatism, faults that you attribute to the binoculars may in fact be caused by your own eyesight. However, if there is a range of instruments available at different price levels you can test a sample from a leading manufacturer, which is unlikely to have the same faults as a cheap model.

Begin by looking at a high-contrast object—a television aerial, branch, or flagpole seen against the sky are favorites. Look for false color around the edges, then move the same object to the edge of the field of view to see whether it remains in focus. If you have to refocus on the object when it is at the edge of the field, you have evidence for a curved field. If no amount of refocusing will make it sharp, the edge distortions are significant. Few binoculars have perfect definition from edge to edge of the field, so don't dismiss the instrument out of hand if this is the case, but decide whether it will be a drawback when viewing a star field. You should expect to get reasonable definition across the central two thirds of the field without having to refocus.

Should you be fortunate enough to be able to observe the Moon, you can really begin to assess the quality of the image. Astronomical objects can often be a severe test of optical quality. The high contrast of the bright Moon will show up any stray reflections in the image, caused by light being reflected off internal surfaces. By day, every part of the image is usually bright, so particular reflections become merged with the rest of the view. But if the dark sky around the Moon has a flare at

Fujinon 16 × 70 FMT-SX Porro

Weight	2170 g 76.5 oz)
True field	4.0°
	70 m/1000 m
	210 ft/1000 yd
Exit pupil	4.4 mm (0.17 in)
Price code	X

Comments

Waterproof. Rugged and with a high reputation, also suitable for marine use.

some point which changes position relative to the Moon as you move it around the field of view, you can be sure that it is a defect of the instrument. In the absence of the Moon any bright light source (other, of course, than the Sun which you should never observe with binoculars) might do.

Look at a repeating pattern such as bricks or railings to check for distortions in the linearity of the field. Pan the binoculars across the pattern to check for a varying magnification across the field.

Any hint of double vision, so that you don't get a restful image, means instant rejection. Test also for undue rocking of the bridge that holds the eyepieces, and ease of focus. Make sure the focus remains constant when you point the binoculars upward.

If you see a fault which the sales assistant blames on the fact that the demonstration model gets a lot of rough use, wonder whether this can be any worse than your own continued use over many years. Any well-built item should be robust enough to withstand being picked up and used. Also bear in mind that sales assistants may not have as much experience in using binoculars as you might expect.

Other checks to make are that the exit pupils appear round and not square, which would suggest that the effective aperture is being reduced by the prisms, and, particularly if you are buying second-hand, that when you look directly into the objectives there are no blemishes or dust on the internal components. This can occasionally be so bad as to give a blurry blob in one half of the glasses when you view a plain area such as sky. It is always well worth looking through more expensive instruments than you plan to buy, so that you have a yardstick against which to measure your choice. Better

Vixen 20 × 80 BCF Porro

Weight	2390 g (84.5 oz)
True field	3.5°
	61 m/1000 m
	183 ft/1000 yd
Exit pupil	4.0 mm (0.16 in)
Price code	F

Comments

Good quality instrument from a leading optical manufacturer. Well suited to deep sky observing.

quality instruments generally have a wider, brighter, and sometimes sharper field of view.

Having selected a particular model, either buy the one you tested or, if you are sold a new boxed unit, subject that to the same tests before parting with your money or at least before you leave the premises. There is no guarantee, even with well-known makes, that every unit will be identical.

Revelation 25 × 100

Weight	3800 g (134 oz)
True field	2.9°
	52 m/1000 m
	157 ft/1000 yd
Exit pupil	4.0 mm (0.16 in)
Price code	D

Comments

Waterproof. Inexpensive large binoculars. Individual focus. Also available under other names. Very heavy—large tripod essential.

6 · USING BINOCULARS

Unlike telescopes, which require considerable expertise to use, control, and maintain, binoculars are pretty straightforward to use and require virtually no maintenance. However, there are some specialized ways of using them and tips on how to get the best view, and these are described in this chapter. The crucial first step, is to get fully dark adapted before you start to look for deep sky objects. Bright indoor lights and viewing a television or computer screen are very bad for night vision, so avoid these if you mean to observe. It can take up to a half-hour for your eyes to gain their full sensitivity to light.

Shield your eyes from extraneous light sources. Get into the darkest part of your observing site, as any extra light can be a distraction. If necessary, sling a blanket over a clothesline to give you a little patch of darkness.

Without wishing to sound overbearing, do wrap up well if it is cold outside. Discomfort is a great disincentive to observing. Ski clothing is a good idea, and always wear a hat as a large proportion of heat is lost through your head. If you are lying on a garden chair, place a blanket beneath you.

Adjusting your binoculars

Many people don't know how to adjust binoculars, and just fiddle with them until things seem right. But there is a definite sequence, which is referred to in the leaflet that should come with your binoculars. Follow it, and not only will you get better views, but you may even be able to reset your binoculars to suit your eyes after someone else has used them.

The first step is to set what is called the interpupillary distance—in other words, the distance between your eyes. Different models provide this adjustment in different ways, but typically the barrels are pivoted on a central bar, on which the focusing ring is also located. The two push together and pull apart around this pivot. This adjustment is usually a matter of trial and error, and in some instruments with thick rubber eyecaps you can find that the correct setting is actually rather uncomfortable as they pinch your nose. It is worth taking care with this step. Make sure that each eye can see the whole field of view, so that it is central to the eyepiece, closing each eye alternately if necessary. But when correctly adjusted, the two circles of light should overlap precisely, and you should not see any black shadows caused by looking through the edge rather than the centerline of each optical system.

As well as these bean-shaped shadows, you will find that if you are not looking centrally—technically speaking, if your eye is not on the optical axis of each half—the images will not be perfect. So if you are having trouble with blurred images or colored fringes around what should be dark edges against a bright background, you may have this adjustment wrong.

Some binoculars have a small scale on the central pivot, sometimes labeled from about 60 to 66. This is the interpupillary distance in millimeters, and it is very useful when sharing one instrument between several people. Get to know your own value and you can reset them quickly. This scale seems to be present less and less on modern binoculars, which is a shame.

The next step is to look through the half of the binoculars that does not have a separate eyepiece adjustment—called the diopter adjustment in the manufacturer's leaflet. Usually, one of the eyepieces can be focused separately by twisting it. The range of adjustment is not great, but is designed to allow for differences in long or short sight between your individual eyes. Ideally you should put a cap on the half that has this adjustment (usually but by no means always the right half), but closing that eye will also do though to a lesser extent. Now choose a suitable object—it doesn't matter how far away—and focus on it carefully with the center focus adjuster using only the one eye.

Finally, cap the half you have just used and look through the other half. This time, look at the same object and focus only using the diopter adjustment on the eyepiece. You should find that you can bring the object to sharp focus. Now the binoculars are correctly adjusted and you can focus on any object, near or far, knowing that any differences between your two eyes are catered for. Sometimes there is a scale of diopters on the eyepiece, ranging from –3 to +3, so again make a note of your personal setting which should be the same whatever binoculars you use.

Occasionally, both eyepieces are adjustable with no separate central focuser, while fixed-focus binoculars may have no adjustment at all, which can make them difficult to use if your eyes deviate much from a standard setting. If you find that the range just does not allow for the differences between your eyes, you probably wear spectacles in everyday life and will need binoculars that you can use while wearing your spectacles. See page 181 for more on this.

Actually achieving the correct focus position calls for a little care. As you know from experience, focusing your eyes on very close objects can be uncomfortable, and it is most restful to view an object that is far rather than near, even if your eyes cannot focus on infinity without spectacles. You can focus binoculars so that the image they give, even of a distant object, is quite close, so your eyes strain to bring it into

◄ *Step 1: Adjust the interpupillary distance by twisting the halves together or apart.*

◄ *Step 2: Using the left half only, focus on an object using the center focus wheel.*

◄ *Step 3: Finally, focus using the right half only using the diopter correcter.*

focus. You can experiment with this: as you turn the focus wheel while looking at a particular object, you will find that you can still keep it in focus through the accommodation of your eye, as it is known. You will find that the best focus position for any particular object is where the eyepiece in question is as far from the objective as possible while giving a sharp image. You can probably feel the eyepiece moving in and out as you focus though some binoculars have internal focusing where this doesn't occur.

Choose a focus position on the outer side of focus rather than the inner side and you will get the most comfortable view without eye strain. Even if you wear spectacles for long or short sight, most binoculars should have a wide enough focusing range to be able to give you perfectly focused views without your spectacles, though some people do find that they need to choose a model with a particularly wide range to cope with their own eyesight. Those who need their spectacles for astigmatism may need to wear them when observing. Most binoculars have rubber eyecaps which fold back on themselves so that you can press your spectacles right up against them.

Supporting binoculars

Binoculars are essentially hand-held instruments, but there comes a point when even the lightest and most convenient instrument becomes a little tedious to hold up to the sky, while for the larger instruments a support of some sort is essential. And although there are of course image-stabilized binoculars, few of us can hold even these rock-steady. The view always suffers a little from the jitters, so again some means of support is a good idea. If you are searching for a particular faint object, or want to study something in detail, binoculars that remain fixed when looking at the object are a blessing.

A garden chair with arms is one possibility, though when viewing objects at any altitude you get very little support from the arms. People often find themselves trying to hold themselves steady against a wall or resting the binoculars on a fence, but this is usually uncomfortable for any length of time.

The obvious solution is a tripod, which is probably the most common means of supporting binoculars. The standard tripod has an adjustable head—either a ball-and-socket, which will tilt in any direction, or a pan-and-tilt, which has separate movements for up-and-down and side-to-side. The ball-and-socket type typically has a single adjuster to vary the friction of the movement between being completely loose and completely firm. It is usually possible to find a happy medium where you can move the binoculars around without too much effort, yet they remain still while you look through them. The pan-and-tilt movements

◀ *The tripod bush is usually concealed behind a cap at the end of the pivot.*

are controlled by levers, which you usually twist so as to increase the friction on that particular axis. Again, you can often find the ideal adjustment point for your purpose.

Either type can be used for supporting binoculars, but if you are buying a tripod specifically for the purpose make sure that the platform that carries the binoculars can be adjusted to tilt fully vertically, that the platform is not too large to fit between the barrels of your binoculars, and that the adjustment lever of a pan-and-tilt head will not get in your way when you are observing, though you may need to reverse the binoculars on the platform so that the lever points away from you. This is not the place to start giving tips on buying tripods, but heed just one caution—the very cheapest and lightest tripods are invariably too wobbly to be of much use, and a cheap pan-and-tilt head in particular is a menace as it will refuse to allow smooth movements. Spend just that little bit more to get something that will be worth using and can also be used for holding a camera or camcorder for more professional results. Large binoculars weigh far more than the average camera, and will need a particularly beefy tripod. Do not underestimate the problem of supporting binoculars that weigh as much as a small bowling ball, and which you need to point very precisely at a small object in the sky. An ordinary tripod will not be nearly good enough.

But fixing the binoculars to the tripod is not as straightforward as with a camera. Tripods are almost always fitted with a threaded stud of a ¼-inch diameter with 20 threads per inch, or what was originally called "quarter Whitworth" after the pioneering engineer Joseph Whitworth who first introduced engineering standards about 1841. Tripods for larger cameras have a ⅜-inch, 16-thread-per-inch stud. These are the threads that are found on cameras worldwide.

However, look at most binoculars and you will not see a bush of this thread. So what's the secret? On many binoculars the bush or socket is at the objective end of the pivot, and is often concealed by a cap,

► *An inexpensive plastic adaptor for fixing binoculars to a tripod.*

occasionally marked with a tripod symbol but more often than not just left blank or with a logo. Sometimes this cap does not even have a knurled edge and is very tricky to remove, particularly on a cold night with numb fingers. It may just pull off. And of course it is a prime candidate for getting lost. Then again, some caps can't be removed at all or after some effort you discover that there is no tripod bush underneath in any case.

Now even when there is a bush, it is at right angles to the normal tripod platform that supports a camera, and indeed the gap between the objectives is too narrow to allow room for the platform. So you need an adaptor, such as the one shown in the picture above, this need not cost much, though the cheapest ones are only plastic and may not last very long. Some manufacturers may have their own proprietary adaptors, which may be expensive, and it may be worth checking on the situation before you buy a particular instrument. Some binoculars just cannot be mounted on a tripod.

For binoculars which have no built-in provision for tripod mounting but have a conventional pivot or shaft, you can buy a different type of

▼ *Russian 20 × 60 binoculars lack a tripod bush.*

▼ *An adaptor such as this fits over the pivot shaft.*

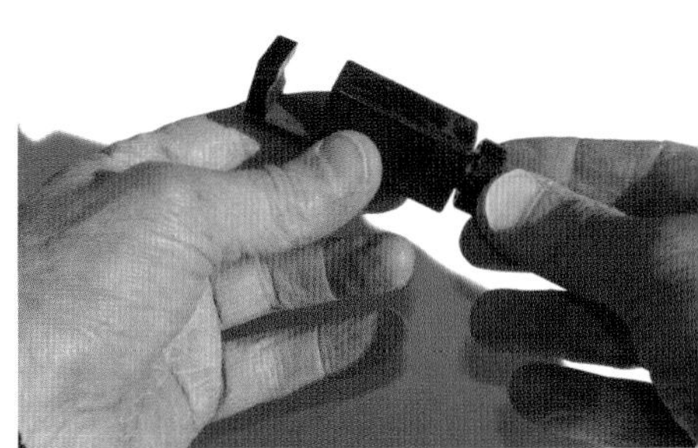

▲ *20 × 60 binoculars on a ball-and-socket head. This has the advantage that there are no long levers to get in your way when viewing.*

adaptor which consists of a post that clamps round the pivot. This works well as long as the pivot shaft is not too wide, and it is accessible—some are not as they are occupied by the focusing mechanism.

Tripod mounting may seem like the perfect solution—it is comparatively cheap and very stable, and the tripod has other uses as well. However, in practice a tripod is not the ideal way to mount binoculars. For one thing, many tripods are not tall enough to allow you to view comfortably when standing; and when you are seated, the legs get in your way. If you want to view an object high in the sky, you have to be uncomfortably close to the tripod.

Another drawback is that when you are viewing, ideally you want to be able to keep your head fixed and move the binoculars, as you do when hand-holding them. But the opposite applies when the binoculars are mounted on a tripod. To look at an object 10° to the left, you have to shuffle 10° to the right. To look at an object 10° higher in the sky, the eyepieces move downward and at a steeper angle. This is particularly awkward if you are observing from a garden chair, for example.

There are both expensive and cheap alternatives to this drawback. The cheap alternative, suitable for binoculars of up to 60 mm, is to use a monopod. This is like a single leg of a tripod, with adjustable height, and essentially with an adjustable head like a tripod. While it does not hold the binoculars or a camera as steadily as a tripod, it does

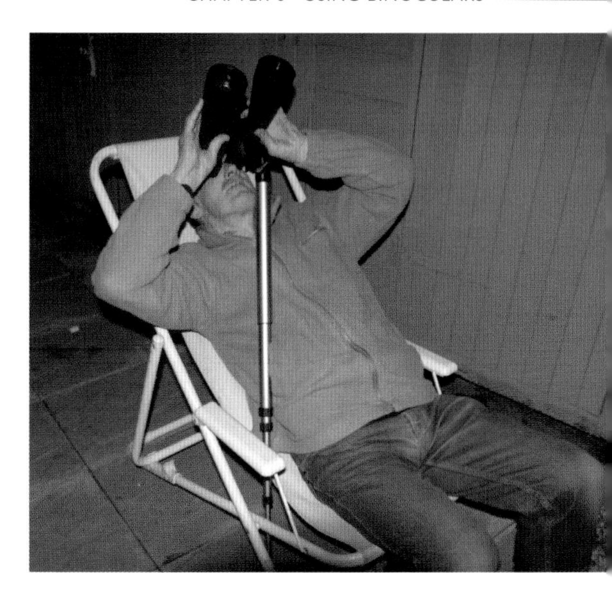

▶ *A monopod helps to support binoculars so you can observe at high elevations in comfort.*

take a lot of the weight and removes a lot of the trembling. And you can move its resting point on the ground very easily, or use that point as a pivot, making it easier to pan around at least part of the sky. Again, it is not usually tall enough to allow you to observe standing up, but if you are seated it can be very helpful. If you are buying one for binoculars, make sure you get one with a fully adjustable head that will allow you to view vertically. Author David Frydman has encountered one model with a rubber adjustment system at the top, which is virtually useless for binoculars.

▶ *Even using a sturdy tripod, observing with large binoculars (here, Revelation 25 × 100s) at high elevations can be uncomfortable.*

◀ Very large binoculars (here, Vixen 25 × 125s) may have angled eyepieces, which allow a much more convenient observing position. This purpose-designed tripod head includes a counterweight.

The expensive alternative is a parallelogram-type mount on top of the tripod, which uses an ingenious system that maintains the viewing angle of the binoculars as you move them up and down. This means that if you leave the binoculars pointing at a particular object, a person of a different height can then see the same object even though they have moved the instrument to their own height. The long extension of the arm also means that the tripod is some way from you, making it easier to lie in a garden chair while observing. But inevitably, the cost of such a system may well be more than you paid for the binoculars. A counterweight is also necessary. There are home-build plans available on the Internet, as the principle is not complicated. People have devised their own systems for mounting binoculars, such as a frame that rests on your shoulders, and if you are of a practical turn of mind you can probably make something for yourself.

A third alternative, providing probably the most comfortable of all viewing positions, is the mirror mount. The binoculars point down at a mirror, ideally fixed to the binoculars themselves, so you can view looking downward while seated at a garden table. You still need to shift the whole unit or your seating position if you want to pan to left or right,

▶ *The Sky Window by Trico Machine allows you to view the sky at a comfortable angle, though the image you see is inverted*

but panning up and down is a matter of tilting the mirror. The mirror itself must be front-coated to avoid double reflections, but need not be of high optical flatness, so this, too, could be a home-build project. There are two drawbacks. One is that the view is laterally inverted, making it harder to locate objects in the first place. The other is that the mirror tends to attract dew even more than the objectives of the binoculars.

Finally, if you have driven into the country to find dark skies, your automobile can provide considerable support for binoculars. Resting them on a half-open window or the door frame, or on the edge of a sunroof, can sometimes be quite convenient.

▲ *The Orion Paragon-Plus is an example of a parallelogram mount, which holds the binoculars at any angle well away from the tripod legs.*

Care of binoculars

If you look after them well, good binoculars will last you a lifetime with virtually no maintenance as long as you replace the lens caps whenever they are not in use. But don't put the caps over dewed-up lenses as the moisture will not have a chance to evaporate. Allow the dew to evaporate naturally. If you have to bring the binoculars in from the cold night air, have a clean plastic bag handy so that they can warm up slowly without being subjected to house dust.

The most common problem is that the objectives become dusty over a period of time. Resist the temptation to try to clean off every speck that falls on them—you will probably do more harm than good. The more you clean the surface, the greater the risk of scratching the coating. What looks like an innocuous speck of fluff could actually be a piece of sharp-edged grit. Use a jet of air to shift dust, but be careful with cans of pressurized air as they can release a jet of propellant as well as the air. A photographic blower brush is a good and safe way to keep lenses clear of loose dust.

Eventually, you may find that there is just so much dust on the objectives that blowing it away does not shift it. You can try using lens-cleaning fluid from a camera shop, applied with a very soft cloth or with lens-cleaning tissues from the same place. There are also proprietary lens-cleaning cloths on the market which your camera shop may stock.

If the moving parts start to sound rough and gritty, be very careful not to probe around and push whatever is there farther into the mechanism. Cleaning these parts is usually best left to a professional repairer.

Collimation

You will, of course, be very careful with your binoculars and would not dream of subjecting them to rough treatment. But occasionally accidents do happen. Good binoculars should survive an accidental bang against the automobile door as you are getting out, but if you do manage to knock the optics out of line, the safest course of action is to take them to a professional repairer, who will put them on a test rig and recollimate them for you.

But sometimes the binoculars are worth less than the cost of a repair. Many years ago, author Robin Scagell managed to drop Russian 12 × 40 binoculars more than three feet onto a concrete path. These inexpensive but well-made glasses suffered a dent in one edge and gave slight double vision, which was easily corrected by a procedure which we offer as a "kill or cure" technique, but which is worth investigating if the binoculars are not very expensive in the first place.

The objectives in binoculars are usually mounted in cells with a deliberately eccentric edge, so rotating them in their cells has the effect of moving the objective slightly. The cells are held in place by metal rings with two slots in which you can insert a suitably shaped home-made piece of metal or even, at a pinch, the ends of a pair of scissors. Undoing the ring and rotating one objective is enough to correct slight double vision as long as the prisms themselves have not been dislodged from their seatings. You try this at your own risk, but it has been known to work on more than one occasion.

Dealing with dew

One of the penalties of observing outdoors is that optical surfaces often dew up, sometimes within minutes. This happens when the temperature of the glass surface drops below what is called the dewpoint, which varies from night to night. Observing the night sky is the worst activity of all from this point of view, as the glass is open to a wide expanse of sky. Even if the air temperature is mild, the temperature of the sky is inevitably much lower, and the heat in the glass radiates away rapidly. The eyepieces also rapidly mist up with the moisture from your eyes.

If dew does form, resist the temptation to wipe it off, which will in due course result in the coating becoming scratched. The only solution is to apply gentle warmth from a hair dryer or radiator. But taking cold binoculars into a warm kitchen usually means that even more moisture will condense, so use the hair dryer outside. Keep a lookout for a portable dryer that will run off an automobile battery as an alternative to a household voltage model.

Telescope owners restrict the amount of sky that their lenses see by attaching what are called dew shields or caps (also sold as lens shades) to the front of their instruments. These are simply tubes of such a diameter and length that they cut down the exposure of the lens to the sky without restricting the field of view of the telescope. But binoculars have a wider field of view, so dew shields are of less use. While 50 mm or 60 mm telescopes are equipped with dew shields as a matter of course, they are virtually unobtainable for binoculars. Fortunately, it is not hard to make your own, so don't throw away any paperboard or plastic tubes of suitable dimensions without testing them for fit on the ends of your binoculars. They will help considerably, and will also help to prevent lights from shining directly onto the lenses, which reduces the contrast of your view.

Telescope owners also use anti-dew heaters, which are low-voltage heating coils run off a special power supply. They are supplied in the form of tape which you can fix round the inside of the dew shields or surrounding the eyepiece, and can be adapted for use with binoculars though the associated wiring robs you of some flexibility when viewing.

Astronomy from indoors

There are several disadvantages to indoor astronomy. If you observe through glass, particularly at high magnifications, you can lose fine definition. You cannot in any case observe at a shallow angle to the glass (rather than at right angles to it) because you will inevitably get ghost images and additional distortions as a result of the greater thickness of glass in the light path. Few of us are lucky enough to have skylights at a range of angles. But if you open a window, there is a risk that warm air inside the house will pour out into the cold night air, causing the image to ripple. These are all reasons why amateur astronomers with telescopes always observe outdoors.

However, with low-power binoculars these problems are not so serious and indeed there are many advantages. First, by observing almost perpendicularly to the glass you can use magnifications of up to 20× or even 30× with little distortion or loss of detail, even with insulated glazing. It is even possible to rest the front of the binoculars against the glass, preferably using binoculars with rubber surrounds to the front barrels. This reduces your hand-shake by about half, and may well allow you to see fainter stars than you would have done outside.

In addition to gently placing both barrels against the glass, you can rest just the left or right side of the binoculars against the window frame. This gives very good stability and using image-stabilized binoculars this way can give a rock-steady image as good as that from tripod-mounted binoculars.

You can observe all manner of objects in this way—the Moon, Jupiter's moons, comets, open star clusters, globular star clusters—and all from a warm and comfortable environment. But you must always be aware of the possibility of ghost images and distortion. If you think you have found a comet with binoculars from indoors, it is more than likely that you haven't— it is probably a ghost image of a bright source either within the field of view, or outside, sometimes well outside, the field. Having said that, the late George Alcock, of Peterborough, England, actually discovered comet IRAS-Araki-Alcock in 1983 through insulated glazing, as well as his sixth nova, in 1991. A nova is a rare type of variable star that suddenly brightens by many thousands of times for a brief period, so an otherwise faint star may become visible to the naked eye.

But in case you think that you might do the same on your first few nights of observing, bear in mind that Alcock was a highly experienced observer who had spent many hundreds of hours scanning the skies for new objects, and knew the sky intimately, so he could spot even a faint interloper immediately.

Observing the Sun with binoculars

As we have already said several times, it is just plain dangerous to observe the Sun directly with binoculars, as permanent eye damage is almost inevitable. But fortunately there are ways and means which are more or less safe, though you still need to use common sense and care.

The simplest method, though one which is going out of favor, is to project the Sun's image onto a screen of some sort. This can be nothing more than a piece of white paperboard or paper. It helps enormously if the binoculars are supported, such as on a tripod. For reasons of safety some people suggest that you cover one half of the binoculars with a lens cap, but this is not a major issue.

Point the binoculars at the Sun, but do not under any circumstances put your eye anywhere near the eyepieces, and do not leave them unattended in this position. The best way to align them on the Sun is to look at their shadow. Hold your screen a distance of about 50 cm (20 in) from the eyepieces to start with, and you should see a bright circle projected through the binoculars onto the screen. This will probably be

▲ *One advantage of projecting the Sun's image is that several people can view it at once, as during the partial stages of this solar eclipse.*

▲ *Solar observing with binoculars using full-aperture filters over the objectives. Take great care that the filters cannot be dislodged.*

blurred, so focus the binoculars until you get a sharp image, which is the Sun itself. You may be tempted to think that this is just a disc of light shining through the circular lenses, but as long as the screen is some distance from the eyepieces it will be the actual Sun.

Look closer and you may see some details. The Sun is not a uniform white disc, but is slightly yellower and darker around its edge, which is known to astronomers as the limb. If clouds cross the Sun, you will see the leading edge sharp and with detail. And if there are sunspots, you will see these as well.

Bear in mind that with this method your binoculars, and in particular the eyepieces, are now subject to quite intense heating, so it is possible to damage them through prolonged use. Cheap binoculars may well have plastic components inside the eyepieces which will easily melt as the Sun's image drifts out of the field of view. The clear cement within the eyepieces may also suffer. If you have ever used a magnifying glass to burn a hole in a piece of paper you will realize that the objectives are focusing the Sun's image in the same way. But with well-constructed binoculars, a brief observing session should do no harm.

▶ *The Coronado Binomite 10 × 25s have solar filters coated onto the objectives for solar observing with no risk of the filters being dislodged.*

The other way to observe the Sun is to cover the objective lenses—never the eyepieces!—with reflective or dense film made especially for the purpose so that you can look through the binoculars directly. It is vital that you use only the approved material, such as Baader AstroSolar film—anything else could actually be more dangerous than not using any filter at all. This is because if a filter looks dark, yet allows infrared through, your eye could be burned even though the Sun looks dim. Normally, a reflex reaction prevents you from staring at anything bright.

However, these solar filters are not sold widely. Only a few specialist suppliers make them. It is possible to make your own quite cheaply from a sheet of the material using spare lens caps that fit perfectly, but you must make sure that there is no possibility that the caps can get dislodged. Some proprietary makes are secured using three set screws.

Coronado, who specialize in solar observation telescopes, make binoculars specifically for solar observation. Binomite binoculars are available in either 10 × 25 or 12 × 60 size, and have solar filters deposited directly onto the objectives giving a neutral image color. Despite the fact that you can view only one object with them—the Sun—devotees of solar work find them great for a quick look to see what the Sun is doing today. The 10 × 25s do not have a tripod adaptor, but the 12 × 60s do.

Photographing through binoculars

The easy answer to this is "don't bother," as there are better ways to photograph astronomical objects. An ordinary telephoto lens has much the same performance as binoculars, and is designed to be fitted to a camera anyway. That glorious star field that you see with your eyes is in reality quite faint, and will require a long exposure time,

▲ *The Moon photographed through 10 × 50 binoculars using an inexpensive digital compact camera. Both binoculars and camera were on a tripod. Photographing at twilight helps to prevent overexposure.*

which in turn means that you will need to track it carefully as the sky rotates. This requires precision astronomical mountings, and is beyond the scope of this book. However, you can have a go at photographing the brighter objects such as the Moon simply by holding a camera up to the eyepiece of your binoculars, which should be held steady in some way.

This is a hit-or-miss operation with compact film cameras, as the main problems are getting the two optical systems precisely in line and in focus. But with digital cameras you can usually see what is going on and can focus much more readily. Even so, you have to get the lens of your camera very close to the eyepiece of the binoculars, which is tricky in itself; and as soon as you have everything in line, you find that the object you are photographing has moved as a result of the daily rotation of the sky. More often than not, you get just a small part of the full field of view of the binoculars. An additional difficulty is getting the exposure right. The autoexposure on digital cameras works well for a normal landscape, but not for a small bright object against a black background, such as the Moon or a planet. The result is an overexposed blob. You really need to be able to control the exposure manually.

Even so, it is possible to get nice photos of the Moon by simply hand-holding the camera up to the eyepiece, and people are often very proud of their results, even if the focus may leave something to be desired. But beyond that, most photos of astronomical objects are doomed to failure.

There are some binoculars that include digital cameras as part of their construction. They usually do not actually take the picture through the binoculars, but instead have a separate long-focus lens for the camera. The only object that is really bright enough to be photographed in this way is the Moon, and it does not seem worth buying such an instrument specifically for astronomy though if the idea appeals to you for general purposes, all the usual comments about testing it for optical satisfaction apply.

Making sketches

It may seem quaintly old-fashioned, in this day and age of sensitive digital cameras, to talk about making sketches of what you see, but there is a long tradition of doing this and it has several advantages. For one thing, you are not limited by the sensitivity of a camera. Although photography in general has a pivotal role to play in astronomy, in most cases this means long-exposure photography, where the light has to fall on the

▶ The globular cluster M53, drawn by Michael Hezzlewood from Burnley, Lancashire, UK, using 12 × 60 binoculars.

sensor for minutes or hours. Only then can a camera exceed the sensitivity of the eye. When it comes to seeing things in an instant, the eye has unrivaled adaptability and sensitivity. No camera can take an instantaneous shot anything like as detailed and with as much sensitivity as your eye can reveal. You can glimpse in a moment what a camera would take many seconds to pick up.

Other advantages of making sketches include that you can build up your own record of what you have seen for future reference. It can be hard to recall just what a particular object looked like, but if you have even a quick sketch you will be reminded very quickly. The discipline of making a sketch forces you to look at the object more carefully, and it forms a basis for comparing future observations, particularly from different locations. It curtails the "seen that, let's move onto the next object" approach which can often leave you unsatisfied by your night's observing, and making sketches can be rewarding. It tests your observing skills, and teaches you to use the techniques of averted vision and so on that actually help you to see more. And it gives you something to compare with the work of others.

No particular artistic ability is needed to sketch most deep sky objects, though it does help to take a little care, and your skill will undoubtedly improve as you go along. It helps enormously to have the binoculars mounted in some way, and to have a convenient and comfortable observing position, which is where the mirror mounted binoculars are a boon. For sketching materials a sketchbook or just plain white paper on a clipboard is adequate, and the type of pencil you use is a matter of preference. Some people prefer a #3 as well as a #2, as it allows greater sensitivity of stroke, but the main thing is to make sure that it is sharp before starting. The most crucial piece of technology is to use a red light. White light destroys your dark adaptation instantly, whereas red light has much less effect and you can return to the eyepiece with little loss of sensitivity.

If you are sketching a deep sky object, start by plotting the positions of the brighter stars in the field of view to act as a basis for positions. Sketching is invariably carried out in negative form, with black stars and dark objects. Use a wide area of the paper rather than trying to cram everything into a little area. If you change your mind over the position of a star, rather than try to erase it on the spot simply put a neat cross through it otherwise it could look like a faint star, or draw a little arrow to show where it should really go. Next, fill in the fainter stars and patterns of stars, using smaller dots for fainter stars. You need not go right to the edge of the field of view—it is usual to show only the center of the field, and even though people might draw a circle around the result this does not necessarily represent the full field of view. Sometimes, people will use a computer mapping program to print out a general view of the brighter stars in a field, then fill in the fainter stars and the object itself using this grid. The drawing of Comet Swan and M13 on page 140 was made using this method, and it has a wider field of view than the binoculars would give at one glance. There is nothing wrong with this as a general technique, and indeed it helps to avoid the overall distortion of scale that can easily creep up on you if you start at the center and work outward. Representing a hazy deep-sky object is achieved by using light shading and then smudging the result with your finger. A #2 pencil is good for this.

It is usually the case that people use the sketches made at the eyepiece as the basis for a finished version that they can complete to a higher quality later on, so make notes at the time to help you in the final drawing. Having completed your sketch, you can scan it and reverse it in a photo editing application to give a more realistic appearance.

Drawing the Moon is a lot more demanding, as there are just so many features to be seen. Use the same general principles of drawing in a general outline then filling in the details. There are many variations of light and shade to be seen, and people sometimes simply draw in the areas with hard outlines and make notes of the brightness for the final sketch.

How faint can you see?

The faintest star visible with any astronomical instrument depends on the aperture and the magnification. Most binoculars have rather lower magnifications than the optimum for seeing faint objects, so they won't show as faint stars as are theoretically possible with such an aperture. The faintest star you can see also depends very much on the observer as well as the observing conditions. Haze, light pollution, and moonlight can have a drastic effect on the faintest magnitude you can reach.

The table here gives a rough indication of the faintest star that you can expect to see from an average rural site using various binoculars, assuming they are tripod-mounted. If you are hand-holding the binoculars you may lose more than a magnitude on these figures, and from a suburban area you will probably lose an additional 1 to 1.5 magnitudes as a result of light pollution. Bear in mind

Binoculars	Faintest star visible
8 × 25	8.8
8 × 40	9.4
7 × 50	9.5
10 × 50	10.0
12 × 50	10.2
15 × 70 or 20 × 60	11.0
20 × 80	11.4

that deep sky objects can be much more difficult to find than their magnitudes would suggest if conditions are poor. You may be able to see stars of ninth or tenth magnitude using a particular instrument from your location, for example, but a galaxy or planetary nebula which is listed as having the same magnitude will probably be much more of a challenge as its light is spread over an area of sky rather than being a point. Binoculars with a higher magnification may work better than those with larger aperture.

Estimating your field of view

You often need to know the field of view of your binoculars, particularly when star-hopping from one known star to a fainter object. The figure may be given on the instrument, but the diagrams shown here give the distances between some easily located stars, and will help you to estimate your field of view at various times of the year.

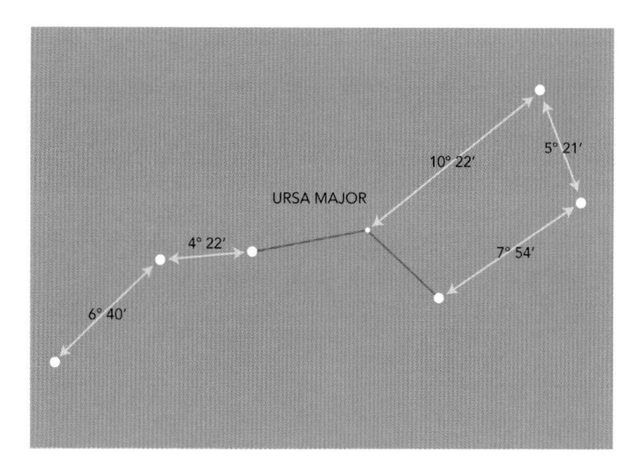

◀ *Ursa Major, visible in the northern hemisphere all year round.*

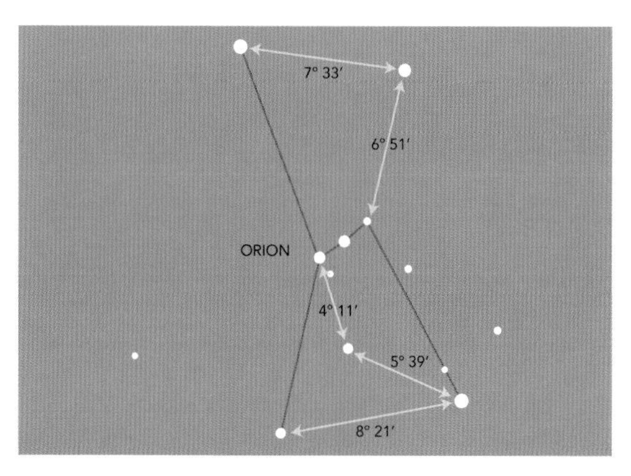

◀ *Orion, visible from everywhere October to March.*

► *Gemini, visible from everywhere December to March.*

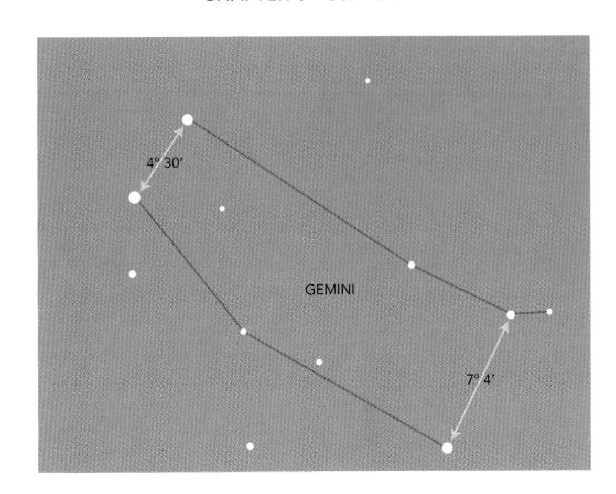

► *Centaurus and Crux, visible in the southern hemisphere all year round.*

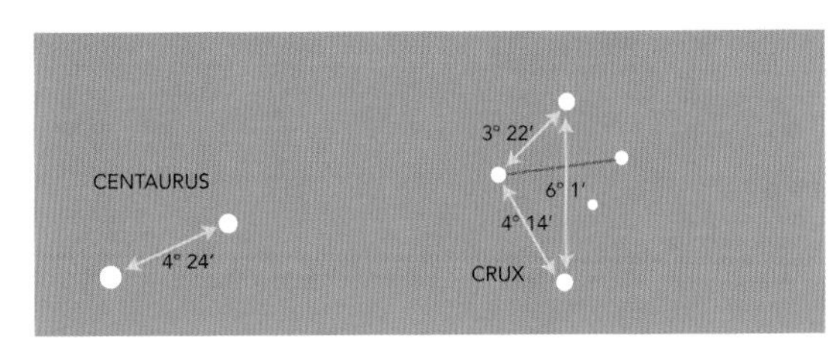

► *Corvus, visible from everywhere March to June.*

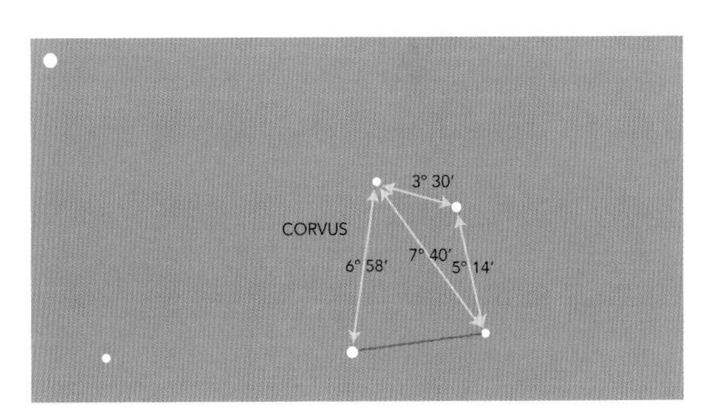

► *Sagittarius, visible from everywhere July to September.*

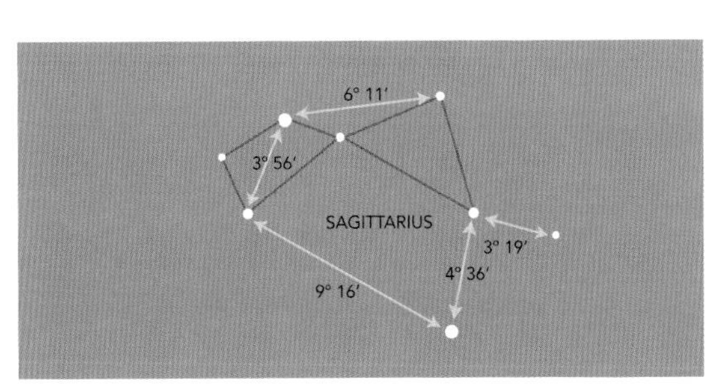

APPENDIX

CONSTELLATIONS			
Name	**Genitive**	**Abbreviation**	**Common name**
Andromeda	Andromedae	And	Andromeda
Antlia	Antliae	Ant	Air Pump
Apus	Apodis	Aps	Bird of Paradise
Aquarius	Aquarii	Aqr	Water Bearer
Aquila	Aquilae	Aql	Eagle
Ara	Arae	Ara	Altar
Aries	Arietis	Ari	Ram
Auriga	Aurigae	Aur	Charioteer
Boötes	Boötis	Boo	Herdsman
Caelum	Caeli	Cae	Chisel
Camelopardalis	Camelopardalis	Cam	Giraffe
Cancer	Cancri	Cnc	Crab
Canes Venatici	Canum Venaticorum	CVn	Hunting Dogs
Canis Major	Canis Majoris	CMa	Great Dog
Canis Minor	Canis Minoris	CMi	Little Dog
Capricornus	Capricorni	Cap	Sea Goat
Carina	Carinae	Car	Keel (of a ship)
Cassiopeia	Cassiopeiae	Cas	Cassiopeia
Centaurus	Centauri	Cen	Centaur
Cepheus	Cephei	Cep	Cepheus
Cetus	Ceti	Cet	Whale
Chamaeleon	Chamaeleontis	Cha	Chameleon
Circinus	Circini	Cir	Compass
Columba	Columbae	Col	Dove
Coma Berenices	Comae Berenices	Com	Berenice's Hair
Corona Australis	Coronae Australis	CrA	Southern Crown
Corona Borealis	Coronae Borealis	CrB	Northern Crown
Corvus	Corvi	Crv	Crow
Crater	Crateris	Crt	Cup
Crux	Crucis	Cru	Southern Cross
Cygnus	Cygni	Cyg	Swan
Delphinus	Delphini	Del	Dolphin
Dorado	Doradus	Dor	Dorado
Draco	Draconis	Dra	Dragon
Equuleus	Equulei	Equ	Foal
Eridanus	Eridani	Eri	River Eridanus
Fornax	Fornacis	For	Furnace
Gemini	Geminorum	Gem	Twins
Grus	Gruis	Gru	Crane
Hercules	Herculis	Her	Hercules
Horologium	Horologii	Hor	Pendulum Clock
Hydra	Hydrae	Hya	Water Snake
Hydrus	Hydri	Hyi	Lesser Water Snake
Indus	Indi	Ind	Indian
Lacerta	Lacertae	Lac	Lizard

CONSTELLATIONS (cont.)			
Name	**Genitive**	**Abbreviation**	**Common name**
Leo	Leonis	Leo	Lion
Leo Minor	Leonis Minoris	LMi	Little Lion
Lepus	Leporis	Lep	Hare
Libra	Librae	Lib	Scales
Lupus	Lupi	Lup	Wolf
Lynx	Lyncis	Lyn	Lynx
Lyra	Lyrae	Lyr	Lyre
Mensa	Mensae	Men	Table (Mountain)
Microscopium	Microscopii	Mic	Microscope
Monoceros	Monocerotis	Mon	Unicorn
Musca	Muscae	Mus	Fly
Norma	Normae	Nor	Level (square)
Octans	Octantis	Oct	Octant
Ophiuchus	Ophiuchi	Oph	Serpent Bearer
Orion	Orionis	Ori	Orion
Pavo	Pavonis	Pav	Peacock
Pegasus	Pegasi	Peg	Pegasus (winged horse)
Perseus	Persei	Per	Perseus
Phoenix	Phoenicis	Phe	Phoenix
Pictor	Pictoris	Pic	Easel
Pisces	Piscium	Psc	Fishes
Piscis Austrinus	Piscis Austrini	PsA	Southern Fish
Puppis	Puppis	Pup	Stern (of a ship)
Pyxis	Pyxidis	Pyx	Compass
Reticulum	Reticuli	Ret	Net
Sagitta	Sagittae	Sge	Arrow
Sagittarius	Sagittarii	Sgr	Archer
Scorpius	Scorpii	Sco	Scorpion
Sculptor	Sculptoris	Scl	Sculptor
Scutum	Scuti	Sct	Shield
Serpens	Serpentis	Ser	Serpent
Serpens caput			Serpent's head
Serpens cauda			Serpent's tail
Sextans	Sextantis	Sex	Sextant
Taurus	Tauri	Tau	Bull
Telescopium	Telescopii	Tel	Telescope
Triangulum	Trianguli	Tri	Triangle
Triangulum Australe	Trianguli Australis	TrA	Southern Triangle
Tucana	Tucanae	Tuc	Toucan
Ursa Major	Ursae Majoris	UMa	Great Bear
Ursa Minor	Ursae Minoris	UMi	Little Bear
Vela	Velorum	Vel	Sails (of a ship)
Virgo	Virginis	Vir	Virgin
Volans	Volantis	Vol	Flying Fish
Vulpecula	Vulpeculae	Vul	Little Fox

GLOSSARY

altitude In astronomy, the distance of an object in degrees above the horizon. Also called *elevation.*

angular measure Means of measuring the separation between objects seen at a distance. There are 360° in a full circle, 60 arc minutes in a degree, and 60 arc seconds in one arc minute.

aperture The clear diameter of an objective lens.

apparent field of view The diameter of the field stop as seen through an eyepiece, measured in degrees. Typically in the range 40°–65°.

aspheric lens elements Lenses with complex curved surfaces, rather than spherical surfaces. More expensive to produce, but can give improved optical performance.

astigmatism An optical error in binoculars or the eye caused by differing magnification at certain orientations.

averted vision Viewing using non-central parts of the eye that are more sensitive to faint light.

BaK4 A type of glass used particularly in binocular prisms for improved optical performance with reduced size and weight.

barrel This contains the glass elements in each side of the binoculars.

barrel distortion A form of distortion of an image in which the magnification at the center is greater than that at the edges, so a square shape will appear to bulge outward.

blooming An older term for lens *coating.*

cell A container for optical components.

cement Transparent medium for joining lens elements to reduce light loss between the adjoining surfaces. May deteriorate over time, leading to reduction in contrast of images.

chromatic aberration The false color, most visible around the edges of objects, introduced into an image as a result of poorly corrected lenses.

close focus The closest distance that binoculars can clearly focus with normal eyesight. Binoculars for nature use may have a close focus of 2 m (6 ft) or less; in standard binoculars it is usually around 5 m (16 ft); with large binoculars it may be as much as 20 m (66 ft).

coating or antireflective coating A material, such as magnesium fluoride or calcium fluoride, deposited onto glass in very thin layers as a means of reducing light loss at the air–glass interface.

collimation In binoculars, the parallel nature of both optical axes so that the images produced by each half overlap exactly when you look through them.

coma An imperfection in an optical system, that results in star images some way from the center of the image appearing not as points but with short tails pointing away from the center.

curved field A field of view in binoculars which does not all come to the same focus position, thereby necessitating constant refocusing, even on sky objects which are all at the same distance.

diopter correction The ability to focus one eyepiece of binoculars separately from the other, thus allowing for correction of the difference in strength between your two eyes. Technically, a diopter is the reciprocal of the focal length of a lens in meters; so a lens of focal length 500 mm (20 in) is 2 diopters.

double vision Occurs when the two halves of binoculars are not exactly parallel, giving images that do not overlap exactly.

ED, extra-low dispersion A type of glass, such as fluorite crystal, used to make lenses which are highly corrected for *chromatic aberration*.

elevation In astronomy, the distance of an object in degrees above the horizon. Also called *altitude*.

exit pupil The diameter of the beam of light that emerges from each eyepiece of binoculars, theoretically given by the diameter of the objective lens divided by the magnification of the eyepiece.

eye cup Shield against stray light at the eye end of an eyepiece.

eye lens The rear lens of an eyepiece, the one which is closest to the eye.

eyepiece The back components of one barrel of binoculars. This examines the image produced by the *objective lens* and provides magnification and field size. Usually consists of three to five, occasionally six, glass elements.

eye relief The distance in mm from the rear glass surface of the eyepiece to the position where the eye sees the whole field of view. With very short eye relief—less than 10 mm (0.4 in)—even non spectacle wearers may not see the whole field.

field glasses Non-prismatic Galilean-type binoculars, typically made in the late nineteenth century. Usually very small field—less than half of prismatic binoculars. Typically 3–5× magnification.

field lens The lens of an eyepiece closest to the front of the binoculars.

field of view The whole scene visible through binoculars, which may be either the *apparent field* or the *true field*.

field stop An aperture in the eyepiece assembly that provides the sharp edge of the field of view of the eyepiece.

fluorite lens A lens that includes one element made from fluorite crystal which makes possibly very good correction for *chromatic aberration*.

focal length The distance between a lens and the image it provides of an object at infinity.

focal ratio The ratio between a lens's focal length and its clear diameter, usually written as, for example, *f*/4 for a lens whose focal length is four times its diameter.

interpupillary distance The distance between the pupils of a person's eyes, usually measured in mm. An average value is 66 mm (2.6 in).
lanyard Strap for binoculars.
light year The distance traveled by light in one year in a vacuum. Equivalent to 9.46 trillion km (almost 5.88 trillion mi).
limiting magnitude The faintest star visible, either to the naked eye or through a given instrument.
LP filter A filter designed to reduce the effect of light pollution—light in the sky scattered from artificial lighting.
magnification The linear increase in apparent size of an object as viewed through an optical instrument. Sometimes (and misleadingly) referred to in advertisements in terms of area or percentage magnification in order to produce larger numbers, where area magnification is the square of the linear magnification.
magnitude A measure of brightness of an astronomical object. A sixth-magnitude star is 100 times fainter than a first-magnitude star, and sixth-magnitude stars are the faintest normally visible from a dark location with the naked eye.
meridian The north–south line at a location, on the ground or in the sky.
monoculars Half-binoculars with a focusing arrangement to correct for eye variations and distance.
multicoating *Coatings* in several layers to improve light transmission across an air–glass surface.
nitrogen or argon sealed Inert gases used within binoculars to prevent internal condensation.
object glass, objective, objective lens The front lens of each barrel of binoculars—usually consisting of two glass elements, sometimes three.
opera glasses Similar to *field glasses* but much smaller and simpler construction. Often found in theaters or as toy binoculars.
optical axis The central line of an optical system, along which any distortions are at a minimum.
optical element An individual optical component in a system, such as the separate parts which together make up a complete objective or eyepiece.
optical group Two or more lens elements cemented together. An eyepiece may have, say, four lens elements in three groups if one pair of elements is cemented together.
Pechan prism A prism system with two separate elements, which inverts a light beam. Two faces of the design have a reflective coating.
pincushion distortion A form of distortion of an image in which the magnification at the edges is greater than that at the center, so a square shape will appear to bulge inward.

Porro prism A prism with one right angled corner and two 45° corners, used in binoculars, which has the property that light shining through the long side will be totally internally reflected out through the same face without requiring a reflective coating. A pair of Porro prisms will invert an image and also shorten the overall light path.

power The same as magnification.

prism A solid glass component with flat surfaces that bend the light, shorten the binoculars and provide the upright image.

prismatic binoculars Any binoculars using prisms to fold and invert the light beam within them, so as to reduce bulk.

resolving power, resolution The separation, in arc minutes or seconds, of the closest double star of equally bright components that can be seen as separate stars using an optical system or the eye.

roof prism A solid prism with a 90° corner, like a roof, which has the property that light entering through one face will be totally internally reflected and inverted by the roof-shaped surface without requiring a reflective coating. A form of roof prism is often used in binoculars.

rubber coating or armor coating Outer layer to the shell of some binoculars, intended to provide impact resistance and waterproofing though often purely cosmetic.

spotting scope A small telescope with an upright image (unlike most astronomical telescopes which have upsidedown images). May have interchangeable eyepieces or a zoom eyepiece. Typically with magnifications in the range 15–60×.

tripod bush Many modern binoculars have a standard ¼-in tripod socket for attaching the binoculars to a tripod sometimes with an adaptor. Some Russian binoculars and spotting scopes have the older ⅜-in sockets.

true field of view The diameter of the scene viewed through binoculars, measured either in degrees or in the number of meters seen at 1000 meters or feet seen at a distance of 1000 yards. Typically in the range 4°–7°.

vignetting Reduction in the brightness of the edge of the field of view as a result of poor design, which results in less than the full aperture being used across the whole field of view.

waterproof Indicates that binoculars can be briefly immersed in water or, in the best cases, that they are genuinely waterproof to a depth of 5 or 10 m (16.5 or 33 ft).

water resistant Questionable term indicating some resistance to rain or damp conditions.

zenith The point directly above the observer, at 90° altitude.

INDEX

Page numbers in *italics* refer to picture captions

CREDITS

Unless listed below, pictures in this book are taken by Robin Scagell/Galaxy Picture Library.

8 tr Nick King/Galaxy
15 David Cortner/Galaxy
18 Manchester Astronomical Society
20-43 Star maps based on Stellarium software
45 t Michael Hezzlewood
51 Philip Perkins/Galaxy
57 Jeremy Perez
60 b Adam Block/NOAO/AURA/NSF/Galaxy
62 t, b Chris Picking/Galaxy
65 bl Michael Stecker/Galaxy
br NOAO/AURA/NSF/Galaxy
67 Yoji Hirose/Galaxy
71 Jeremy Perez
75 Bill Schoening/Todd Boroson/NOAO/AURA/NSF/Galaxy; ESO (NGC 5128)
77 Chris Picking/Galaxy
81 STScI/AURA/Galaxy
82 Eddie Guscott/Galaxy
85 Michael Stecker/Galaxy
90 Eddie Horsley
93 NOAO/AURA/NSF/Galaxy
96 Michael Hezzlewood
105 t Michael Hezzlewood
b Bill Schoening, Vanessa Harvey/REU program/NOAO/AURA/NSF/Galaxy
107 Chris Picking/Galaxy
115 Michael Stecker/Galaxy
116-7 NASA/Galaxy
118 Michael Rosolina
124-5 James Jefferson/Galaxy
128 Juan Carlos Casado/Galaxy
135 STScI/Galaxy
140 Jeremy Perez
144 Jonathan Bell
145 Jonathan Bell
159 Canon UK
160 Olympus UK
161 Bushnell Outdoor Products
163 Swarovski
164 b Swift Sport Optics
166 Carl Zeiss Sports Optics
167 b Pentax UK
168 Celestron
169 Steiner
170 Optical Vision
171 Bresser Optics
172 Leica Camera Ltd
173 Nikon UK
174 Celestron
175 Fujinon Europe Gmbl
176 Vixen Optics
187 t Trico Machine
187 b Orion Telescopes & Binoculars
191 Keith Sugden/Gal.
192 Paul Sutherland/Ga.
193 Steve Collingwood/Telescope House
195 Michael Hezzlewood

Our thanks to Telescope House for the loan of Revelation binoculars.